Praise for *5 Principles* *Modern Mathematics Classroom* by Gerald Aungst

"Math is linear . . . Math is neat and tidy . . . Math is about getting the right answer . . . Not in this book! Gerald Aungst has done an awesome job of showing us what math learning can look like in a way that embraces the messy inquiry and problem solving that accompanies real-world math. Put this book to work and watch your students be amazing!"

—Scott McLeod, Director of Innovation, Prairie Lakes Area Education Agency, and Founding Director, UCEA Center for the Advanced Study of Technology Leadership in Education

"Bravo! *5 Principles of the Modern Mathematics Classroom* brings a conceptual framework for K–12 mathematics to life. As a parent of a struggling math student and as the executive director of Edutopia, a source for what works in education, I commend Gerald Aungst for sharing his 5 Principles of the modern mathematics classroom. This book is a perfect blend of inspiring and practical. Highly recommended!"

—Cindy Johanson, Executive Director, Edutopia, George Lucas Educational Foundation

"Gerald Aungst has written a book about math instruction and learning, but the strategies can easily trickle to other disciplines in school. He not only shares great ideas about how we can create meaningful learning opportunities and environments for students, he clearly articulates why this is important. As someone who is not a math teacher, I found what Aungst wrote to be applicable to much of the work that I am doing in all areas, and really felt that the environment of math learning that he shared is much better than the one I experienced as a student."

—George Couros, Innovative Teaching, Learning, and Leadership Consultant

"Gerald Aungst ignites the magic of mathematics for his readers through 5 beautifully articulated guiding principles. Reminding us of what makes mathematicians so passionate about their subject matter, Aungst not only grounds his work in research but also takes us on a journey into several classrooms in each vignette so that we may take away actual, practical tips to put into practice today."

—Erin Klein, Teacher, Speaker, and Author of *Redesigning Learning Spaces*

"Gerald Aungst reframes mathematical thinking for the reader, giving space for teachers to go beyond the usual airtight methods for math pedagogy."

—Jose Luis Vilson, Educator and Author of *This Is Not a Test*

"This is the book that teachers have been waiting for—evidence-based practices that are easy to implement. Aungst walks us through simple ways to change our mindset about math, encourage students to embrace mathematical challenges and integrate technology in our daily practices. So many ideas for every grade level!"

—**Anne M. Beninghof**, Teacher, Speaker, and Author

"In this book, Gerald Aungst expertly weaves math and literacy to promote a culture of thoughtful inquiry, problem solving, and innovative thinking. He consistently demonstrates deep respect for the topic, the teachers, and the students, carefully interspersing stories, strategies, questions, tips, and tools in just the right places throughout each chapter to maximize understanding using a gradual release of responsibility model. My eyes are forever opened to an instructional world where math and literacy not only coexist—but work in concert to elevate our instructional power potential!"

—**Dr. Mary Howard**, Author of *Good to Great Teaching*

"A dynamic read for math educators K–12, Aungst challenges the status quo and many traditional methods of classroom instruction that are still prevalent in many of our nation's classrooms. In doing so, he makes a case for the vitality of a student-centered, rich classroom culture, filled with collaboration, problem-solving, rigorous curriculum, and high-quality communication, while celebrating student success at every turn. Moving from theory to practice, and providing a myriad of concrete methods and examples, *5 Principals of the Modern Mathematics Classroom* is a must-have tool for every math teacher's professional toolbox!"

—**Thomas C. Murray**, State and District Digital Learning Director, Alliance for Excellent Education, Washington, DC and author of *Leading Professional Learning*

"Take a good hard look at your teaching practices! If your math classroom resembles your childhood classroom, then you need to invite Gerald Aungst to the party. His five practices will help you transform your teaching so you can develop the critical thinkers of tomorrow. Gone are the days when mathematics instruction was focused solely on procedures. As Aungst describes so astutely in terms of music, we must train our students to be ready to work through unique problems just as we train music students to sight read. He provides sound research and examples from all grade levels, including websites and apps for incorporating technology. The book addresses the pressures of state tests and the long-term benefits of an actively-engaged classroom. Aungst supports the premise that ALL students can learn math. The book will guide you along with reflective questions and suggestions for how to begin. This is an inspirational must-read, and includes insightful questions for use in a book study. Put it on your reading list and begin the journey."

—**Mary Palladino**, Math Coach

"Gerald Aungst is as innovative as he is refreshing in his approach to mathematics instruction. The strategies he describes will make an average teacher excellent and an excellent teacher outstanding. The examples in this book can be used immediately in 1-on-1 conversations with teachers, small group professional development, or large group workshops. Each chapter resonates with his approach to the modern mathematics classroom. This book is an excellent resource for every teacher who teaches mathematics."

—**Sean W. Gardiner**, EdD, Director of STEM Education, Upper Merion Area School District, PA

"This book is great for our educational world right now. There have been many who are talking about the changes in our math classrooms to help promote more problem solving and not just answering questions. This is a great start for teachers to begin to make changes in their classrooms for students to become great problem solvers."

—**Anthony Purcell**, Sixth Grade Math Teacher,
Longfellow Middle School, Enid, OK

"Gerald Aungst masterfully reminds us that math education must be more than a 'recipe' and challenges us to aid our students in thinking like mathematicians. With practical advice and inspiration, you'll again find the beauty in math and be empowered to reveal it for your students. Regardless of the age you teach, this book offers a fantastic breakdown of the role of conjecture, collaboration, communication, chaos, and celebration play in transforming your classroom into one where students engage free-range thinking every day."

—**Kelly Tenkely**, Architect of Learning, Founder and
Administrator, Anastasis Academy, Centennial, CO

"This book is practical and forward thinking. It should be required reading for undergraduate college math methods classes."

—**Christine Ruder**, Third Grade Teacher,
Truman Elementary School, Rolla, MO

"I absolutely recommend this book! Elementary teachers are overwhelmed with how to teach multi-step problems to young children. Knowing they (the teachers) don't have to be the Know-All of the class, but to allow the students the opportunity to think, possibly fail, repeat, provides the teacher time to watch the students and help direct their learning."

—**Lyneille Meza**, Coordinator of Data and Assessment, Denton ISD, Denton, TX

"Math is an area in which critical thinking skills are vital. This book will assist educators in encouraging students to think critically when dealing with math problems. Learning to incorporate the 5Cs into the math classroom is sure to enhance both teaching and learning! . . . Our school always welcomes material that will enhance students' ability to learn and develop critical thinking skills! (And even have some fun while learning!!!)"

—**Susan E. Schipper**, First Grade Teacher, Charles Street School, Palmyra, NJ

"*5 Principles of the Modern Mathematics Classroom* is an easy-to-read and use book that provides educators with practical strategies and solutions to immediately incorporate into their mathematics classrooms. The design of the book provides classroom teachers an opportunity for deep reflection of their current mathematics instructional practices to ensure that they are creating students who are mathematical thinkers and problem solvers. The book offers teachers strategies to allow their students to wonder freely, work collaboratively with their peers, communicate efficiently, and participate in experimentation and celebration. Using technology as a way to incorporate the principles of a modern mathematics classroom is briefly explored with short explanations of specific tools for each principle. Overall, Aungst has provided his readers with a 5-principle framework that when incorporated into a classroom, will transform students into mathematicians."

—**Stephanie Schwab**, STEM Program Administrator,
Montgomery County Intermediate Unit, Norristown, PA

"What does it take to be an excellent teacher? With so many ideas about teaching floating about, it is necessary to capture the essential components and package them for careful consideration. While perspectives for what it actually takes to ensure all students flourish are manifold, it is certain that the aspiring teacher must be constantly reflecting on her practice and expanding her awareness of productive beliefs and actions. To this end, Gerald Aungst has developed a guide ideal for any teacher hoping to find the sweet spot between theory and practice that can easily translate to the classroom. Simple and engaging, the stories and resources inspire the reader to delve into an entirely unplanned investigation of new ideas at every turn. The *5 Principles of the Modern Mathematics Classroom* are engaging, perplexing, and ripe for exploration and discussion. I'd recommend the read for any teacher looking to better understand what being an excellent teacher means through the lens of the classroom culture and for any PLC looking for a head start in casting a vision for a strong mathematics culture in their school."

—Levi Patrick, Director of Secondary Mathematics
Education at the Oklahoma State Department of
Education, Co-Founder of OKMathTeachers.com

"In *5 Principles of the Modern Mathematics Classroom: Creating a Culture of Innovative Thinking*, Gerald Aungst comprehensively describes the characteristics of an environment for mathematical learning which integrates the math practice standards and actively involves students in higher level, inquiry-based thinking. Aungst's book holds value for the beginning teacher, who will find a multitude of resources and ideas for establishing an effective mathematics classroom, and for the seasoned teacher, who will be challenged to reflect on her own practice and incorporate the 5 Principles into her math instruction. Coaches and other instructional leaders will find the book well-suited for book study or PLC discussion, with contemplative questions at the conclusion of each chapter. The 5 Principles provide a framework for propelling math learning forward in the twenty-first century."

—Louise Kirsh, Math Coach,
North Penn School District, Landsdale, PA

"Mathematical thinking critically impacts a learner's capability to process the world around her. The typical math instruction actually degrades a child's capability to see math in the world and to find delight in the messy construction of analytical, critical, and creative processes through mathematical thinking. In *5 Principles of the Modern Math Classroom*, Gerald Aungst provides a learning path for teachers that leads them to think together and to practice math instruction that makes math real. And that's what great math teachers do–at any age."

—Pam Moran, Superintendent,
Albemarle County Public Schools, VA

"One of the major strengths is that the book is written in teacher friendly language and that it gives practical examples. This content is much needed for today's teachers who are either stuck in a traditional way of teaching OR wanting to change and are looking for ideas."

—Barbara Fox, Adjunct Faculty, Lesley University,
University of Boston, Regis College, Boston, MA

5 Principles of the Modern Mathematics Classroom

Creating a Culture of Innovative Thinking

Gerald Aungst

Foreword by Mark Barnes

FOR INFORMATION:

Corwin

A SAGE Company

2455 Teller Road

Thousand Oaks, California 91320

(800) 233-9936

www.corwin.com

SAGE Publications Ltd.

1 Oliver's Yard

55 City Road

London EC1Y 1SP

United Kingdom

SAGE Publications India Pvt. Ltd.

B 1/I 1 Mohan Cooperative Industrial Area

Mathura Road, New Delhi 110 044

India

SAGE Publications Asia-Pacific Pte. Ltd.

3 Church Street

#10-04 Samsung Hub

Singapore 049483

Acquisitions Editor: Erin Null

Signing Editor: Desirée A. Bartlett

Editorial Assistant: Andrew Olson

Production Editor: Melanie Birdsall

Copy Editor: Janet Ford

Typesetter: C&M Digitals (P) Ltd.

Proofreader: Jennifer Grubba

Indexer: Maria Sosnowski

Cover Designer: Candice Harman

Marketing Manager: Rebecca Eaton

Photographs by Gerald Aungst.

Original artwork created by Jerald Gottesman, http://jginkcreative.com

Printed in the United States of America

Library of Congress Cataloging-in-Publication Data

Names: Aungst, Gerald.

Title: 5 principles of the modern mathematics classroom: creating a culture of innovative thinking/Gerald Aungst; foreword by Mark Barnes.

Other titles: Five principles of the modern mathematics classroom

Description: Thousand Oaks, California: Corwin, [2016] | Includes bibliographical references and index.

Identifiers: LCCN 2015027594 | ISBN 9781483391427 (pbk.: alk. paper)

Subjects: LCSH: Mathematics—Study and teaching. | Motivation in education.

Classification: LCC QA11.2 .A96 2016 | DDC 510.71—dc23 LC record available at http://lccn.loc.gov/2015027594

This book is printed on acid-free paper.

Certified Chain of Custody
Promoting Sustainable Forestry
www.sfiprogram.org
SFI-01268

SFI label applies to text stock

15 16 17 18 19 10 9 8 7 6 5 4 3 2 1

Contents

To Michele for your support and patience while I pursue my passions.

Foreword

I was undeniably one of the most inept math students ever to open a textbook with numbers scattered across the cover. My fondest memory about me and math—an odd couple for sure—is receiving a graded test from my teacher in ninth grade. She called attention to my high score, and I smiled inwardly, embarrassed at being singled out from the crowd. You see, I'd never done well on anything math related in any of my prior school years. "Obviously, he cheated," one mischievous peer shouted from the other side of the classroom, garnering a chorus of giggles from most of my other upstart friends. "Oh, I don't know," Mrs. Wilson said. "He may have stumbled on it." She glanced back at me for only a second and nodded, as if to say, "Nice job, kid. I know you did honest work." I managed a C+ in the class. For my part, I felt neither pride nor frustration over the grade. I do recall, though, thinking that I'd learned something that year, and I'll never forget Mrs. Wilson, the first teacher who ever made me believe I could provide a cogent response to a question about x or y.

The next year brought a different math focus, a new teacher, and the all-too-familiar results. I was again a horrible math student, reminded why I hated numbers my whole life before ninth grade. It wasn't until many years later that I realized it wasn't math that I hated; it was math class. Mrs. Wilson had shown me math in a different way. She made sense of it. "I don't believe anyone hates math," Gerald Aungst once told me. "I think people who think they hate math really just hate the way it's taught." A lifetime of failure in math classes, save for that one year with Mrs. Wilson, made me a skeptic, but Aungst was convinced that he could change my mind.

In *5 Principles of the Modern Mathematics Classroom*, Gerald Aungst takes me back more than 35 years to a time when math made sense. This is ironic, since the book is about the principles of modern math, but very early in this book, you'll realize what I realized: The modern math class isn't so much about the 21st century, as it is about a modern approach to math instruction. And while I may be decades removed from Mrs. Wilson's class, I realized while reading this book that what I hated was math classes that were rooted in old-school strategies, classes that instilled a belief that math was nothing more than an endless stream of nonsensical numbers dropped on yellowed worksheets. Is this an impeachment of teachers? I don't think so. Rather, it's an indictment of methods. I remember other math teachers I encountered in my K–12 school years who were friendly and kind and a few who were even funny—a quality I loved in teachers then and still value today. They greeted me at the door, called me by name, and cracked jokes that made students laugh and feel welcomed in their classrooms. Then, these same wonderful people distributed worksheets with 25 problems that made my eyes glassy and sent my mind wandering, as they lectured endlessly about the value of ratio and probability and vocabulary that I was unwilling to learn.

In 13 years, only one teacher encouraged me to think critically about math. She made me want to solve problems. She invited me to ask questions and to seek the answers. She didn't mind when class became a bit chaotic, and she knew when it was important to celebrate success. She believed in me. Her class was not modern, at least in terms of how we typically think of the word, but her practice was modern in the sense that Gerald Aungst presents the word in this book.

Unfortunately, when I became a teacher more than a decade later, I forgot how Mrs. Wilson made me appreciate math class. Instead of giving my students wonderful, rich novel characters and exciting opportunities to explore the wonders of reading and writing, I showered them with workbook activities and multiple-choice tests. They couldn't stand it. Just as I didn't really hate math, they didn't dislike language arts. What my students hated—even those with an affinity for literature—was my class. It took me far too long to learn what Mrs. Wilson knew when I was only 15 and what Gerald Aungst teaches us in this book—how to teach "modern."

Now, nearly a quarter century after I taught my first class, I teach teachers student-centered education, digital learning strategies, and how to assess learning without numbers and letters (i.e., traditional grades). The teachers I encounter in the field and on social media typically have the best intentions for their own students. Educators inherently want to help kids. I do find many math teachers, though, who struggle with the idea of teaching modern. "How can students learn math without repeated practice?" some ask, while others wonder what is to be learned in a classroom rife with noise, movement, and what most onlookers perceive to be chaos. For these teachers and all others interested in captivating young people with fractions, decimals, ratios, and other potentially frightening symbols and words related to mathematics, it's time to go modern, and no one does this better than Gerald Aungst.

I met Gerald a long time ago at an education conference and immediately knew he was the kind of teacher I admire. We talked about involving students in the learning process and how important it is to give them choice in the activities that help demonstrate mastery. He alluded to my work, emphasizing the value of inspiring students to participate in the assessment process. Not long after that initial meeting, someone in a Facebook group I moderate asked a question about progressive, student-centered learning in a math class, and Gerald instantly came to mind. I knew he would answer the question much better than I could. Gerald replied to the math teacher's inquiry, speaking eloquently about the power of collaboration, communication, and chaos—words I've used often in my writing and keynote speeches, but wasn't accustomed to hearing about math instruction.

How does one know if children understand problems without the use of traditional tests? How can students collaborate on basic math? What alternatives are there to grades, when 2 + 2 always equals 4? These issues continue to plague math teachers and while I want to tell them to eliminate rote memory work and change the ways they assess, I struggle to provide other options. Now I realize the answer to all of these questions is in the *5 Principles of the Modern Mathematics Classroom* and, surprisingly, the strategies that English, Science, and History teachers use can work equally well in Math class. Gerald Aungst teaches us that problem solving does not occur on worksheets. It requires a different kind of thinking. Problems are solved when students ask more questions than their teachers. Skills are acquired through innovative use of digital tools, through failure, and with iteration. All of this creates a culture of problem solving in math class. Gerald Aungst reveals the beauty of this culture in ways that will have teachers eager to get back to school, so they can modernize their instruction and make even students like me appreciate the beauty of math.

—**Mark Barnes**
Author/Educator
Publisher of the Hack Learning Series
Founder of Teachers Throwing Out Grades

Preface

Mathematics instruction is frequently conducted under a false assumption: that mathematics is a fixed, linear sequence of skills that must be acquired one by one, building block by building block, until students have achieved mastery. This leaves out what makes mathematicians pursue their subject with passion and drive: the mystery and magic of math. The magic is captured in what I believe is the most important part of any mathematics curriculum: the processes, practices, and habits of mind we want to foster in all students. In order to embrace the magic, the classroom must be a place where problem solving happens daily and is deeply embedded in its culture.

> The mystery and magic of math is captured in what I believe is the most important part of any mathematics curriculum: the processes, practices, and habits of mind we want to foster in all students.

WHY I WROTE THIS BOOK

Throughout my career, first as a classroom teacher, then a gifted support teacher, and now as a math supervisor, I've had the opportunity to visit many math classrooms. I have seen many excellent teachers who know math content and pedagogy for teaching the learners in their classrooms, but who still approach mathematics instruction in a mechanical, traditional manner. For many years, I was one of those teachers.

As I learned more about mathematics, and about teaching and learning, my thinking evolved, and I developed or acquired many new strategies for teaching math. I came to understand that the classroom environment forms the foundation for effective math instruction. As I made sense of the various elements of a thinking-oriented classroom, the beginnings of this book took shape.

In writing this book, I hope to share the benefits of my own experiences, research, and mistakes over many years, and to help other teachers understand how to transform their own classrooms.

AUDIENCE

This book is primarily aimed at classroom teachers of mathematics in Grades K–12. Principals and instructional coaches will find it useful in supervising teachers as a guide for what to look for during classroom visits and how to assist teachers in improving their practice. In addition, curriculum directors will find it valuable as a reference when planning curriculum. Finally, college instructors working in preservice education programs will find the book can serve as a primary or supplemental resource in mathematics methods courses.

SPECIAL FEATURES

This book contains a number of special features that help you navigate and process all of the material within.

The 5 Principles

This book reveals 5 Principles of the modern mathematics classroom. Each Principle encompasses several orientations or practices that enable the teacher to build that thinking-oriented culture in the classroom:

 Conjecture. This process entails inquiry, questioning, and problem finding.

 Collaboration. Students primarily work in pairs and groups, supporting, encouraging, and helping each other.

 Communication. The modern mathematics classroom is as much about reading, writing, speaking, and listening as it is about computation. A core element is that students frequently formulate and support mathematical arguments.

 Chaos. Real problem solving, and therefore real math, is messy, and modern mathematics classrooms need to not only tolerate, but support and promote ways to work through the mess.

 Celebration. This step emphasizes effort and learning over right answers and achievement. Students are encouraged to adopt a growth mind-set.

Throughout the book, the icons serve to draw attention to places where specific ideas and strategies connect multiple principles together. A large icon indicates the main Principle associated with the strategy, and smaller icons show additional Principles that the strategy also supports strongly.

Classroom Vignettes

In addition, the book describes specific classroom routines that support each of the principles. These provide techniques a teacher can use to grow a culture over time that enables students to become confident problem solvers. Classroom vignettes illustrate the strategies in practice, with examples both from real classrooms and composite scenarios.

Examples by Grade Band

To show that the Principles apply to any grade level, each chapter includes a section that highlights one of five grade bands: K–3, 2–5, 4–7, 6–9, and 8–12. Each band shows a specific strategy or technique that supports the Principle, and illustrates what that principle looks and sounds like in the classroom. Within each section, one or two specific strategies or techniques are listed that are useful for establishing a problem-solving culture for students in those grades.

These grade bands intentionally overlap to emphasize three points:

- Although a strategy or example is framed for a particular grade band, the Principle behind it applies regardless of grade level.
- Students in a given grade have very different needs and capabilities, and you can adopt a variety of tools and strategies to meet all of those needs.
- All of the strategies in this book should be considered as illustrations of much broader ideas. As such, I encourage the reader to innovate and look for ways to adapt all of the techniques to any specific classroom situation regardless of grade level. Techniques useful in one grade can usually be adapted and applied to another, so read every section with an eye toward crafting your own blended approach.

Technology Integration
Tips and Case Studies

Each chapter also describes digital technology tools which can be integrated to facilitate and support the outlined routines and practices. These

> I encourage the reader to innovate and look for ways to adapt all of the techniques to any specific classroom situation regardless of grade level.

sections are not intended as extended tutorials in each tool—that is beyond the scope of this book. Instead, my purpose is to point you in the direction of a few specific tools that show you some possibilities for application.

I also include several technology case studies that illustrate a few of these tools in practice in real classrooms.

Connections to the Standards for Mathematical Practice

Too often we build our curriculum only on content. But, mathematics educators have advocated teaching processes for decades. Half of the Principles and Standards for School Mathematics from the National Council of Teachers of Mathematics (2000) are process standards. In 2001, the National Research Council published *Adding It Up: Helping Children Learn Mathematics* (Kilpatrick, Swafford, & Findell, 2001), whose five strands of mathematical proficiency are about the processes students should learn. Most recently, the Common Core State Standards have expressed these ideas in the *Standards for Mathematical Practice* (National Governors Association Center for Best Practices, and Council of Chief State School Officers [NGA/CCSSO], 2010). Since these standards are being used by many schools around the country, I reference them throughout this book. However, the 5 Principles are broader and more universal than the Common Core, and are applicable regardless of the foundation of your curriculum design.

It's important to note here that the mathematical practices are included not as a checklist to be completed for each technique, but rather to make visible the relationships between specific practices and the 5 Principles. This can be helpful when you are asked to document how your instruction is meeting standards or promoting connections to the practices.

Book Study Questions

Each chapter concludes with several study questions. These are designed to provoke thinking and reflection about your own work and classroom environment, and to take you beyond the text of the chapter into new ideas and explorations.

If you're like me, you might skip over the questions in order to get on to the next chapter of content. Resist this urge. The questions are an integral part of the content and thinking about them helps you to crystallize your own understanding of the material.

HOW TO USE THIS BOOK

Though you can certainly read the book through from start to finish and get a great deal out of it, I recommend some other approaches that might help you gain more from the experience.

Individual Study. If you are reading this on your own, first read Chapters 1 and 2 thoroughly. Next, quickly read Chapters 3 through 7 to get the big picture of the 5 Principles. Now read Chapter 8 thoroughly to begin learning about how to transform your classroom environment. Next, I recommend you choose one Principle as a focus and reread that chapter in great detail and depth. Be sure to answer the reflection questions, and use the planning tools in Chapter 8 to begin implementing strategies over the course of a month or so. Every three to four weeks, choose another Principle, reread that chapter, and implement the strategies.

PLC/Study Group. If you are using this book as part of a Professional Learning Community (PLC) or study group and plan to study it for a full school year, you could take a similar approach to the individual study above, with the addition of regular group discussions. For a shorter eight to ten week study, you could read a chapter each week and focus your discussions on the book study questions at the end of each chapter. For both approaches, follow each discussion session with practical implementation by visits to each other's classrooms and detailed feedback provided on what you see and hear.

Whole School Faculty or Administrative Team. For a schoolwide or administrative team study, the book can be a good guide for framing professional development activities and developing common language for supervision. I recommend following a similar schedule to one of the PLC studies above during the first year of implementation. Staff members get an overview of the Principles and the techniques. Each following year, one of the Principles can be chosen as an emphasis. Encourage the collection and sharing of new techniques, which over time will build a rich repertoire for supporting thinking in the classroom. Look for applications of the 5 Principles outside of math to strengthen the interdisciplinary connections to other content areas.

> Look for applications of the 5 Principles outside of math to strengthen the interdisciplinary connections to other content areas.

WHAT THIS BOOK IS NOT

It is **not** about mathematics content. Teachers of math at all levels need a strong foundation in the content they teach, however, and elementary teachers in particular need greater depth and skill in math than they are likely to have received in their training. There are a number of excellent books, courses, and resources to improve your skill with math content.

It is not a book of "tips and tricks" for math instruction, nor is it a book of math activities or lesson plans. While there is some value in these, they do little good when the tips and activities are injected in isolation into a classroom environment that isn't designed to support them.

WHAT YOU WILL GET OUT OF THIS BOOK

Although there are a growing number of books about supporting thinking and problem solving in the classroom, this book is uniquely suited to the needs of practicing and preservice educators in the following ways:

- It provides both a theoretical background and solid practical applications that can be used in the classroom today.
- It is applicable no matter what grade level you teach, providing specific illustrations and techniques relevant to your situation.
- It promotes 21st century skills and integration of digital technology with specific tools and techniques to support and enhance the Principles for each grade band.

Acknowledgments

M any people influenced the thinking that led to this book, and many others contributed in ways small and large to the final product. It is a hopeless task to attempt to remember and thank them all, so I will not try.

However, a few people do deserve particular mention.

During the writing process, I constantly picked the brains of my Pennsylvania colleagues in both the School District of Cheltenham Township and the Montgomery County Intermediate Unit. In particular, I appreciate the direct and indirect support and insights of Dwight Nolt, Jeanne Rauch, Matthew Bramucci, Dorie Martin, Stephanie Schwab, and Donna Gaffney.

The acknowledgments in a book on education would not be complete without mentioning teachers. I've had a number of good ones, but there are two, Bud Oehrli and David Ickes, whose influence led directly to this book. From third grade through sixth grade, they nurtured my natural abilities in both math and writing, and modeled many of the strategies I formalized in writing this book.

However, before even those teachers came my parents, Roberta and Lester Aungst. They believed in me and encouraged me to follow my passions, no matter how many times I changed majors in college. They were my first teachers, and they have modeled for me what it means to be a lifelong learner. My father was also my first example of what a professional educator can and should be.

Throughout the process, the staff at Corwin have been extremely supportive and helpful to this first-time author. In particular, editor Desirée Bartlett has redefined patience and graciousness with her rock steady assistance for many, many, many months.

Finally, thanks must go to my three boys, Justin, Timothy, and Daniel, and my beautiful and generous wife Michele. Their love and support kept me going when deadlines loomed and I needed just a couple more hours to write, or when I needed to be reminded of why I do the work that I do. I appreciate you more than you will ever understand.

PUBLISHER'S ACKNOWLEDGMENTS

Corwin would like to thank the following individuals for their editorial insight and guidance:

Roxie R. Ahlbrecht, Mathematics Intervention Specialist/ Educational Consultant
Retired from Sioux Falls Public 49–5

Karen Creech, Fifth Grade Teacher
Creighton's Corner Elementary
Ashburn, VA

Lyneille Meza, Coordinator of Data and Assessment
Denton Independent School District
Denton, TX

Anthony Purcell, Sixth Grade Math Teacher
Longfellow Middle School
Enid, OK

Christine Ruder, Third Grade Teacher
Truman Elementary School
Rolla, MO

Debra A Scarpelli, Math Teacher
Slater Middle School
Pawtucket, RI

Susan E. Schipper, First Grade Teacher
Charles Street School (preK–6)
Palmyra, NJ

Tom Young, Math Teacher
Spring Lake Park Senior High
Spring Lake Park, MN

About the Author

 **Gerald Aungst** is the supervisor of gifted education and elementary mathematics for the School District of Cheltenham Township in Pennsylvania. Prior to his service as an administrator, Gerald taught mathematics at the elementary level for eighteen years in both the regular classroom and as a gifted support specialist. He also ran a summer remedial program for elementary students for four years. Most recently, Gerald has been part of his district administrative team leading the transition to the Pennsylvania Core Standards and training teachers in high quality math instructional practices.

Gerald is a cofounder of the websites All About Explorers (www.allaboutexplorers.com) and Connected Teachers (www.connectedteachers.org).

He has been a member of the Alliance for Excellent Education's Project 24 Team of Experts and the Digital Learning Day planning team. He also served on the national administrator task force for the National Association for Gifted Children (NAGC) and has been a member of the board of directors for the Pennsylvania Association for Gifted Education (PAGE).

Gerald is an experienced speaker, and was a member of the keynote panel and a featured presenter at Digital Learning Day 2013. He has also presented at national and state conferences, including the International Society for Technology in Education (ISTE), Florida Educational Technology Conference (FETC), EduCon, the Pennsylvania Educational Technology Expo and Conference (PETE&C), Iowa Technology and Education Connection (ITEC), National Association for Gifted Children (NAGC), and the Carnegie-Mellon Institute for Talented Elementary and Secondary Students. Gerald is also on the organizing team for Edcamp Philly.

Gerald holds a bachelor's degree from Penn State University and a master of education degree from Widener University. Gerald blogs at his website (www.geraldaungst.com) and is a regular contributor to Brilliant or Insane (www.brilliant-insane.com). His writing has also appeared at Edutopia (www.edutopia.org), in the Association for Supervision and Curriculum Development (ASCD) *Educational Leadership* magazine, and a chapter in the book, *Teaching Literacy in the Digital Age: Inspiration for All Levels and Literacies*. You can find Gerald on Twitter at @geraldaungst.

1

Math for the 21st Century and Beyond

Being a mathematician is no more definable as "knowing" a set of mathematical facts than being a poet is definable as knowing a set of linguistic facts.

—Seymore Papert, MIT Mathematician, Computer Scientist, and Educator

The moment we believe success is determined by an ingrained level of ability, we will be brittle in the face of adversity.

—Josh Waitzkin, American Chess Player, Martial Arts Competitor, and Author

MATHEMATICS AS LITERACY

"But, I'm just not that good at math." How many times have you heard this? It may be said when figuring the tip on a restaurant check, or trying to make sense of some statistic reported on the news. The unstated subtext is, "It's OK if I get this wrong, because I haven't got the innate ability."

This is hardly a new phenomenon. John Allen Paulos, in his classic book, *Innumeracy: Mathematical Illiteracy and Its Consequences,* calls it a "perverse pride in mathematical ignorance" (Paulos, 2001, p. 4). Unlike similar failings, such as struggles with reading or social skills, being bad at math is something people boast about.

> We assume that we are born with whatever mathematical talent we're ever going to have. Math is therefore something we either get or don't get, completely dependent on that talent. I have a word for that: *hogwash*.

We assume that we are born with whatever mathematical talent we're ever going to have. Math is therefore something we either get or don't get, completely dependent on that talent. I have a word for that: *hogwash*. Kimball and Smith (2013) point out that "for high-school math, inborn talent is much less important than hard work, preparation, and self-confidence" (para. 2). Mathematics classrooms are often built to reinforce the former perception instead of the latter truth. "One of the most damaging mathematics myths propagated in classrooms and homes is that math is a gift, that some people are naturally good at math and some are not," says Jo Boaler of Stanford University. "This idea is strangely cherished in the Western world, but virtually absent in Eastern countries, such as China and Japan that top the world in mathematics achievement" (Boaler, 2014, para. 2).

Sousa (2008) points out another important truth, which makes the job of math educators more difficult. "Spoken language and number sense are survival skills; abstract mathematics is not. In elementary schools we present complicated notions and procedures to a brain that was first designed for survival in the African savanna. Human culture and society have changed a lot in the last 5,000 years, but the human brain has not" (p. 1).

Math has acquired a reputation as a specialized and esoteric subject. The thinking goes, if I don't grasp it I'm clearly not one of the elite few who have what it takes. But, people who learn to read late in life are celebrated as heroes. Why shouldn't the same be true for math?

In order for the brain to learn these important skills and concepts, teachers need to create an environment where mathematical reasoning is essential. People learn a language best when immersed in it. Math is no different. We need to immerse students in environments that provide organic opportunities to use and apply math concepts. This book lays out the framework for creating such an environment. To understand that framework, it's important to recognize that mathematics has far more to offer our students than the ability to accurately compute a server's tip.

MATHEMATICS AS AN INNOVATION INCUBATOR

Ask most people today to name an important technology innovator and you are likely to hear names like Steve Jobs, Bill Gates, Salman Khan, and Mark Zuckerberg. Yet, their work would have looked very different without a woman who rarely gets mentioned as a technology innovator. You may not even recognize the name *Grace Hopper*, but the problems she solved paved the way for those future innovators.

Dr. Grace Hopper was a mathematics professor and a U.S. Naval officer. She was also among those who developed the UNIVAC computer in the early 1950s. Her great innovation in the field of programming was the idea of the *compiler*, a program that takes human friendly computer code and translates it into the machine code that computers use.

Prior to Hopper's invention, computer programs were written directly in machine code, a time-consuming and labor-intensive process. Each program was also particular to the machine for which it was written. This meant if you wanted to run a program on a different computer, you had to start again from scratch. Grace Hopper's compiler opened the door for high-level languages. Programmers could now develop code that was independent of the hardware and could be reused on multiple machines.

Although this seems a fairly mundane concept in the mid-21st century, it was a radical idea seventy-five years ago. In fact, even after it was working, Hopper's colleagues dismissed her compiler as impossible. "I had a running compiler and nobody would touch it," she said. "They told me computers could only do arithmetic" (Scheiber, 1987, para. 8). Fortunately for those of us who consider a smartphone to rank near air, water, and food on the "need" scale, they were wrong.

Popular media tell us innovation is essential to the future of our society and economy. But, like math aptitude, we view innovation as something intangible and fleeting, a characteristic of the rare genius. Innovation and creativity *can* be learned, however, and many schools are striving to become places that nurture these characteristics in all children.

Although innovative ideas and practices happen in every field of human endeavor, and therefore in every subject studied in school, I believe the mathematics classroom is the best place to create conditions that enable innovation and creativity to thrive. We do this through the practice of solving problems.

This seems counterintuitive. After all, in math, the answer is always the answer, right? Two plus two always equals four, and no amount of creativity or innovation will change that. So, how can a math class become an environment that encourages those creative and innovative mind-sets?

Consider the following statements—all of which are true, mathematically valid, and will confound most students, if not their teachers:

1. $9 + 5 = 2$

2. $8 + 8 = 10$

3. A triangle can have three right angles

4. $16.75 + 15\% = 20$

5. $375 \div 70 = 6$

One key to understanding these is to realize that the context for the statement matters. For example, in the first, we are talking about time. If you begin at nine o'clock and move ahead five hours, you end at two o'clock. The second occurs often in the context of computer science, and the third involves non-Euclidean geometry.

The last two may seem a bit of a cheat, since the computations are not strictly precise. However, they are both valid statements in real-world situations. It is very likely you used the fourth example within the last week or two: it is a typical example of figuring a server's tip. Rarely do we compute a tip to the penny. More often, we just round to a convenient amount because more precision isn't necessary. The last statement is the mathematical expression of this problem: "375 fourth graders are going on a field trip. Each school bus holds 70 students. How many buses do you need?" If you consider the expression alone, the "right" answer would have a remainder or a fraction. A fraction of a bus makes no sense in this context, however, and therefore is an incorrect answer to the question.

Mathematics, when taught as a framework for problem solving and reasoning rather than as a collection of rules and algorithms to memorize, opens us to flexible ways of thinking about the world. Instead of seeing "solvability" as a fixed property, students need to see *every* problem as potentially solvable, and see the possibilities that could lead them toward a viable solution.

> Mathematics, when taught as a framework for problem solving and reasoning rather than as a collection of rules and algorithms to memorize, opens us to flexible ways of thinking about the world.

This was Dr. Hopper's true legacy. Later in her life, she said "The most important thing I've accomplished, other than building the compiler, is training young people. They come to me, you know, and say, 'Do you think we can do this?' I say, 'Try it.' And I back 'em up. They need that. I keep track of them as they get older and I stir 'em up at intervals so they don't forget to take chances" (Gilbert, 1981, p. 4). Most interesting, perhaps, is that humans are born with this "take chances and try it" mind-set. Why, then, would we need to reinstall it in our students? Two reasons are that we train this adventurousness out of kids at a very young age, and that school culture is designed to reinforce this training.

A study conducted at University of California at Berkeley illustrates what happens. Psychologists Christopher Lucas, Alison Gopnik, and Thomas Griffiths (2010) gave preschoolers and adults similar problems to solve. They had a box which would light up and play music when certain types of objects, which the researchers called "blickets," were placed on top. The task for the participants was to determine which of the objects had the "blicketness" property and would therefore turn on the machine. Sometimes, a single object would turn the box on, and sometimes the objects needed to be placed in combination.

The researchers found that the preschoolers were much more likely than the adults to solve the problem correctly. Even when the solution involved subtle, abstract relationships between the objects, young children were consistently better. One reason, they said, is that children are much more fluid in their thinking and aren't hampered by prior experiences and expectations. They are more willing to try many solution paths, even those that seem unlikely. Whereas adults hone in on one likely option and try it repeatedly, even when it doesn't seem to be working.

> Children are much more fluid in their thinking and aren't hampered by prior experiences and expectations. They are more willing to try many solution paths, even those that seem unlikely.

One explanation for this adult behavior is the way we teach math. We discourage free-range thinking, and instead insist on students learning the "one true way" that a particular kind of problem should be solved. And this restrictive teaching happens very early: the same researchers found that toddlers were much more flexible problem solvers than Kindergarteners. "Our current educational system better prepares children to answer questions that are well defined and presented to them in the classroom than it does to formulate the nature of problems in the first place. Often the skills involved in solving well-defined problems are not the same as those involved in recognizing a nonobvious problem or creating a problem" (Pretz, Naples, & Sternberg, 2003, p. 9).

My explanation for the adults' behavior in this experiment was not considered in the Berkeley study, and to my knowledge is an unanswered research question. There are likely other factors besides traditional methods of math instruction that contribute to the loss of fluid thinking. Regardless, we should be designing school environments to reverse that trend rather than accelerate it.

With the arrival of the Common Core State Standards, schools across the United States are reexamining their mathematics curricula. Although they carry political baggage, the standards give us a tremendous opportunity to think deeply about our practices. If we do nothing else, centering our math instruction around problem solving and the mental habits that go with it will go a long way toward improving learning outcomes for our students.

GOALS OF THIS BOOK

This book aims toward two goals:

1. To provide a framework for K–12 mathematics education designed to create a modern, immersive environment for nurturing the problem-solving skills of all children so they can realize their potential to be the innovators of the next half century.

2. To empower all teachers, regardless of their environment and personal level of comfort with and knowledge of mathematics content, to become stronger teachers of math.

The next chapter further expands on these two ideas, creating a deeper foundation for understanding the 5 Principles framework. I discuss the habits of mind that are necessary for better math learning, the problem with traditional approaches to math teaching, and the meaning and value of "rigor." This sets the stage for understanding why I focus on classroom culture and link the Standards for Mathematical Practice from the Common Core State Standards (CCSS) with the 5 Principles.

WHAT ELSE TO EXPECT

In Chapters 3 through 7, I present each of the Principles for cultivating problem solving and innovative thinking: Conjecture, Collaboration, Communication, Chaos, and Celebration. These Principles form the framework that supports the habits of mind and curriculum goals previously outlined in Chapters 1 and 2.

> This book is about making fundamental changes to the culture of learning that happens throughout the course of an entire year. While you and your students can benefit from putting just a few strategies into place, the greatest impact will come if you take the time for deep reflection and thoughtful redesign of the ways you do business every day.

I caution you to avoid cherry-picking one or two intriguing ideas from a chapter and then stopping. This book is about making fundamental changes to the culture of learning that happens throughout the course of an entire year. While you and your students can benefit from putting just a few strategies into place, the greatest impact will come if you take the time for deep reflection and thoughtful redesign of the ways you do business every day. Take a moment to let that sink in, but don't be intimidated. The process of changing your practice and your classroom culture requires commitment, but it is absolutely within your capability. Plan to spend time with this material. A major theme of this book is that *effective learning* does not necessarily come from *efficient teaching*. Don't sacrifice the former for the latter.

To assist with this, in the final chapter I describe a process for creating and sustaining these changes in your own classroom—first by assessing the current state of your classroom culture, then by designing a plan for implementing this framework in stages, and finally by creating the support system which enables you to sustain the changes you want to see in your classroom.

FOR PLC AND STUDY GROUPS

1. Have you ever said you were "not a math person"? Why do you think this might be? What experiences did you have that made you feel that way? Could a teacher have done or said something that would have changed this feeling?

(Continued)

(Continued)

2. Do you agree with the premise that mathematics is a good place to teach flexibility and innovative thinking? Why or why not? What objections do you have to this approach, and how might that affect the way you teach math?

3. Without knowing the specifics of each Principle, which of them resonates most with you as something you think you might be comfortable with: Conjecture, Communication, Collaboration, Chaos, or Celebration? Why do you think this is so? What connection do you think that comfort level could have with teaching math?

4. Which of the 5 Principles gives you the most anxiety when you think about incorporating it into your math teaching? Why? Based just on the name, what barriers can you identify which you will have to begin breaking down in order to be most successful in implementing that principle?

2 Classroom Environment

A Medium for Change

The world doesn't care what you know. What the world cares about is what you can do with what you know.

—Tony Wagner, Innovation Education Fellow at
the Technology & Entrepreneurship Center at Harvard

You can't be a lifelong learner if you have all the answers.

—Adam Bellow, Educator and
Founder of EduClipper and WeLearnedIt

WHY CLASSROOM CULTURE?

In the last chapter, I cautioned you to avoid using this book as only a collection of tips, tricks, lessons, and project ideas, and instead to see it as an integrated framework for redesigning a classroom culture.

Why is this so important? To find out, let's take a side trip to . . . mushrooms.

Kennett Square, Pennsylvania, is called the "mushroom capital of the world." Chester County, where Kennett Square is located, produces more than a million pounds of mushrooms per day, almost half of the nation's mushroom crop (United States Department of Agriculture, 2013). Once a year, local mushroom farms give tours as part of the annual Mushroom Festival. I recently visited one of those farms, and during my hour-long visit I learned a great deal about what it takes to grow Maitake mushrooms.

A Maitake mushroom farm is a surprisingly simple place. Large cinderblock rooms have several aisles, each lined with long, wooden shelves. These are stacked about two feet apart, rising two stories high. Shelves are lined with row after row of plastic bags containing a mixture of red oak sawdust and a proprietary blend of nutrients. Mushroom spawn are introduced into the substrate, and the bags are left for several weeks in the cool, dark, damp environment to grow. At the farm I visited, each room contains about ten thousand bags, each of which produces about a pound of Maitake mushrooms. Rooms are started about a week apart so that the farm can consistently turn out ten thousand pounds of crop a week.

If I were to pluck a few of the tips I learned during my visit, such as keeping the room at sixty-one degrees Fahrenheit and cutting open only one side of the bag during the last two weeks to promote a larger, tighter head on the end product, I might think that I could easily begin to grow my own mushrooms. In fact, I could have picked up mushroom "kits" that were sold at the Festival for a few dollars to try growing them at home.

However, despite its simple appearance, in reality a commercial mushroom farm is a complex system which must all work well in order to reliably produce crops week after week and year after year.

Creating a modern mathematics classroom based on the 5 Principles that reliably nurtures good problem solvers is no different. For the same reason that I couldn't throw up a couple shelves in my basement and grow high-quality mushrooms, you cannot expect to pick and choose a few examples from this book and then see a significant change in your students' thinking—for a classroom to become more fertile to problem solving, there must be an overhaul of the culture in the classroom.

> For the same reason that I couldn't throw up a couple shelves in my basement and grow high-quality mushrooms, you cannot expect to pick and choose a few examples from this book and then see a significant change in your students' thinking—for a classroom to become more fertile to problem solving, there must be an overhaul of the culture in the classroom.

The 5 Principles aren't just for students who are at present high achievers or for those who have already mastered the basics. A problem-solving approach focused on process instead of content is essential to helping struggling learners as well. Boykin and Noguera (2011), in their book *Creating the Opportunity to Learn*, also list many of these principles in their description of classroom environments which effectively eliminate racial and ethnic achievement gaps. No matter who you teach or where, the 5 Principles framework can help you improve learning for your students.

DEPTH AND RIGOR

Consider the following:

> Miguel collects baseball cards. Last week he had 217 cards in his collection. Today, his aunt gave him two dozen more for his birthday. How many cards does he have now?

This is a common type of problem seen in elementary-level textbooks. You might find it at the end of a worksheet about adding multi-digit numbers. The process for solving is straightforward, though it does require a few mental steps to complete it successfully.

Project Mentoring Mathematical Minds (Project M³) is a research-based program designed to challenge and motivate mathematically talented students in Grades 3–6. Now compare the baseball card problem with this Project M³ problem adapted from Lesson 2, *"Card Game Capers"* from the book *Mystery of the Moli Stone* (Gavin, Chapin, Dailey, & Sheffield, 2006):

> You and your friends are going to play a game using a set of cards numbered from 0 to 9. On your turn, you are going to draw three cards from the facedown deck, one at a time. The object is to make the largest 2–digit number you can using your cards, with the leftover card being discarded. The catch is that you must decide where to write each digit before you draw the next: tens place, ones place, or discard. If you draw a 4 as your first card, where should you write it, and why?

This, too, is a problem, but it seems different in some fundamental ways. As adults, we can quickly see the correct path to the solution of the baseball card problem, and finding that answer is just a matter of working our way through that path. The card game question, on the other hand, feels fuzzier.

The first example was designed to apply one specific, abstract mathematical skill to a concrete situation which might occur in the real world.

The second, however, is designed to apply various ways of reasoning and conceptualizing about numbers and their relationships.

Most of what passes for "problems" in available math resources are of the first variety. They aren't actually problems so much as they are exercises. Wikipedia defines an exercise as "a routine application of . . . mathematics to a stated challenge. Teachers assign mathematical exercises to develop the skills of their students" (Exercise [mathematics], 2014, para. 1). The key word here is *routine*. Like physical and artistic exercises, they are intended to be repeated frequently until a particular process becomes fluent and automatic. Mathematical exercises are the cognitive equivalent of scales in music, ball handling in sports, and knife skills for a chef. Important for the practitioner to do, certainly, but they don't constitute a complete performance.

> Mathematical exercises are the cognitive equivalent of scales in music, ball handling in sports, and knife skills for a chef. Important for the practitioner to do, certainly, but they don't constitute a complete performance.

We may think that it's enough to have students complete more story problems. But, Resnick (1988) points out

> Story problems do not effectively simulate out-of-school contexts in which mathematics is used . . . [T]he language of story problems is highly specialized . . . requiring special linguistic knowledge and distinct effort on the part of the student to build a representation of the situation described. Furthermore, this representation, once built, is a stripped down and highly schematic one that does not share the material and contextual cues of a real situation. (p. 56)

Instead of giving students real problems to solve, we've just given them yet another set of mathematical symbols to manipulate. This is far from the depth of understanding that we want.

The word "rigor" is hard to avoid today, and it provokes strong reactions from educators. Policymakers tout its importance, and publishers promote it as a feature of their materials. But, some teachers share the view of Joanne Yatvin, past president of the National Council for Teachers of English. To them, rigor simply means more work, harder books, and longer school days. "None of these things is what I want for students at any level" (Yatvin, 2012, para. 3).

We have adopted jargon without clearly understanding it. "'People don't know what it means,' said longtime educator and consultant Barbara Blackburn. 'The teachers I work with are being told they're supposed to include rigor. It's certainly the flavor of the month. But teachers all say everyone is telling me what to do but they can't tell me how to do it'" (Colvin & Jacobs, 2010, para 20). For classroom teachers, then, the more important question is one of practice: How do we create rich environments where all students learn at a high level? One useful tool, Norman Webb's (2005) Depth of Knowledge Levels, can help teachers meet that challenge. Depth of Knowledge (DOK) categorizes tasks according to the complexity of thinking required to successfully complete them.

Level 1. Recall and Reproduction. Tasks at this level require recall of facts or rote application of simple procedures. The task does not require any cognitive effort beyond remembering the right response or formula. Copying, computing, defining, and recognizing are typical Level 1 tasks. Recall of basic math facts and application of memorized algorithms would be math tasks at this Level.

Level 2. Skills and Concepts. At this level, a student must make some decisions about his or her approach. Tasks with more than one mental step, such as comparing, organizing, and estimating are usually Level 2. Math examples would include when a student has to select from among several possible well-defined paths or algorithms, or has to use an algorithm in an unconventional, but straightforward way. The baseball card problem in the last section is an example of a Level 2 task, since the student must first determine the numeric value of "two dozen," then select and apply the correct algorithm.

Level 3. Strategic Thinking. At this level of complexity, students must use planning and evidence, and thinking is more abstract. A task with multiple valid responses where students must justify their choices would be Level 3. Examples include designing an experiment, or analyzing characteristics of a genre. In mathematics, solving a non-routine problem, or explaining the reasoning behind a Level 2 application would be examples of Level 3 tasks. The number-card game in the previous section is a good example of a Level 3 task.

Level 4. Extended Thinking. Level 4 tasks require the most complex cognitive effort. Students synthesize information from multiple sources, often over an extended period of time, or transfer knowledge from one domain to solve problems in another. Designing a survey and interpreting the results, analyzing multiple texts to extract themes, or writing an original myth in an ancient style would all be examples of Level 4. A Level 4 math task would involve multiple sources of raw data, or complex problems requiring innovative thinking with no routine solution path.

> DOK levels are not developmental. All students, including the youngest preschoolers, are capable of strategic and extended thinking tasks . . . All students should have opportunities to do complex reasoning with advanced content . . . even Kindergarteners.

You may be asking at this point, "Well, what is a reasonable distribution? How often should I be doing tasks at each level? What's the right sequence?"

DOK levels are not sequential. Students need not fully master content with Level 1 tasks before doing Level 2 tasks. In fact, giving students an intriguing Level 3 task can provide context and motivation for engaging in the more routine learning at Levels 1 and 2.

DOK levels are also not developmental. All students, including the youngest preschoolers, are capable of strategic and extended thinking tasks. What those tasks look like will differ. Tasks that require Level 3 reasoning for a fifth grader may be a Level 1 or 2 task for a high school student who has learned more sophisticated techniques. All students should have opportunities to do complex reasoning with advanced content. Recent research strongly supports this, even for Kindergarteners. "All children, regardless of their early childhood care experiences, benefit from more exposure to advanced mathematics content" (Claessens, Engel, & Curran, 2014, p. 426; see also Engel, Claessens, & Finch, 2013). To find the right balance, ask yourself these questions: "What kinds of thinking do I want students to routinely accomplish? If my own children were participating, what would I want them to be doing? What's the most effective way to spend the limited classroom time I have?" You should decide how often you focus on tasks at each level so students gain the most from the learning opportunities you design.

Regardless of how you define "rigor," the important thing is that students are thinking deeply on a daily basis. Webb's Depth of Knowledge outline gives us a framework and common language to achieve that effect in your classroom.

Distinguishing Between DOK Levels in Math

Math specialist and teacher Robert Kaplinsky has developed a tool to help teachers recognize the differences between Depth of Knowledge levels in mathematics tasks. The tool, available at http://robertkaplinsky .com/tool-to-distinguish-between-depth-of-knowledge-levels/, provides explicit examples of problems for all grade levels illustrating DOK levels 1, 2, and 3. Figure 2.1 gives a selection of these examples. Try each of the problems yourself so you can experience the kind of thinking needed to solve each level. In particular, notice that Levels 2 and 3 require a more sophisticated understanding of the concept and cannot be solved through a rote or routine process.

Now try creating your own set of problems for a topic you are teaching soon. Begin with a DOK 1 exercise like one you might find in your math textbook. Then develop problems of increasing depth that require more complex application. There is no formula for this process: keep referring back to the DOK levels, and share and discuss the problems you've created with colleagues or students.

Figure 2.1 Kaplinsky's (2015) Tool to Distinguish Between DOK Levels

Topic	Area and Perimeter	Probability	Quadratics in Vertex Form
CCSS Standard(s)	3.MD.8 4.MD.3	7.SP.5 7.SP.7	F-IF.7a
DOK 1 Example	Find the perimeter of a rectangle that measures 4 units by 8 units.	What is the probability of rolling a 5 using two standard 6-sided dice?	Find the roots and maximum of the quadratic equation below: $y = 3(x - 4)^2 - 3$
DOK 2 Example	List the measurements of 3 different rectangles that each has a perimeter of 20 units.	What value or values have a $\frac{1}{12}$ probability of being rolled using two standard 6-sided dice?	Create three equations for quadratics in vertex form which have roots 3 and 5, but have different maximum and/or minimum values.
DOK 3 Example	What is the greatest area you can make with a rectangle that has a perimeter of 24 units?	Fill in the blanks to complete this sentence using the whole numbers 1 through 9, no more than one time each: Rolling a sum of ___ on two ___-sided dice is the same probability as rolling a sum of ___ on two ___-sided dice.	Create a quadratic equation using the template below with the largest maximum value using the whole numbers 1 through 9, no more than one time each: $y = -\Box(x - \Box)^2 + \Box$

A FOCUS ON THE PRACTICES

Classrooms all over the United States are now using the Common Core State Standards (CCSS) as the benchmark for the content and skills to teach in mathematics. As districts and schools realign their curricula to these standards, it would be easy to focus on the content standards and use them as a checklist for coverage. This approach misses a critical foundational layer. I believe the most important part of the CCSS for math are the Standards for Mathematical Practice, which define eight ways of thinking and reasoning about mathematical ideas. In this book, I refer to these as "the Practices" or by the initials "MP."

One reason the Practices are often missed is that, despite being placed at the front of the CCSS document, they are not directly embedded into any of the grade-level standards. They are intended to be overarching standards used and taught throughout all grades. Let's take a look at each of the Practices.

MP1: Make sense of problems and persevere in solving them. Mathematically proficient students start by explaining to themselves the meaning of a problem and looking for entry points to its solution. They analyze givens, constraints, relationships, and goals. They make conjectures about the form and meaning of the solution and plan a solution pathway rather than simply jumping into a solution attempt. They monitor and evaluate their progress and change course if necessary. Mathematically proficient students check their answers to problems using a different method, and they continually ask themselves, "Does this make sense?" They can understand the approaches of others to solving complex problems and identify correspondences between different approaches.

MP2: Reason abstractly and quantitatively. Mathematically proficient students make sense of quantities and their relationships in problem situations. They bring two complementary abilities to bear on problems involving quantitative relationships: the ability to decontextualize—to abstract a given situation and represent it symbolically—and the ability to contextualize, to pause as needed during the manipulation process in order to probe into the referents for the symbols involved. Quantitative reasoning entails habits of creating a coherent representation of the problem at hand; considering the units involved; attending to the meaning of quantities, not just how to compute them; and knowing and flexibly using different properties of operations and objects.

MP3: Construct viable arguments and critique the reasoning of others. Mathematically proficient students understand and use stated assumptions, definitions, and previously established results in constructing arguments. They make conjectures and build a logical progression of statements to explore the truth of their conjectures. They justify their conclusions, communicate them to others, and respond to the arguments of others. Mathematically proficient students are also able to compare the effectiveness of two plausible arguments, distinguish correct logic or reasoning from that which is flawed, and—if there is a flaw in an argument—explain what it is. Students can listen or read the arguments of others, decide whether they make sense, and ask useful questions to clarify or improve the arguments.

MP4: Model with mathematics. Mathematically proficient students can apply the mathematics they know to solve problems arising in everyday life, society, and the workplace. In early grades, this might be as simple as writing an addition equation to describe a situation. In middle grades, a student might apply proportional reasoning to plan a school event or analyze a problem in the community. By high school, a student might use geometry to solve a design problem or use a function to describe how one quantity of interest depends on another. They are able to identify important quantities in a practical situation and map their relationships using such tools as diagrams, two-way tables, graphs, flowcharts, and formulas. They can analyze those relationships mathematically to draw conclusions.

MP5: Use appropriate tools strategically. Mathematically proficient students consider the available tools when solving a mathematical problem. These tools might include pencil and paper, concrete models, a ruler, a protractor, a calculator, a spreadsheet, a computer algebra system, a statistical package, or dynamic geometry software. Proficient students are sufficiently familiar with tools appropriate for their grade or course to make sound decisions about when each of these tools might be helpful. When making mathematical models, they know that technology can enable them to visualize the results of varying assumptions, explore consequences, and compare predictions with data. Mathematically proficient students at various grade levels are able to identify relevant external mathematical resources, such as digital content located on a website, and use them to pose or solve problems. They are able to use technological tools to explore and deepen their understanding of concepts.

MP6: Attend to precision. Mathematically proficient students try to communicate precisely to others. They try to use clear definitions in discussion with others and in their own reasoning. They state the meaning of the symbols they choose, including using the equal sign consistently and appropriately. They are careful about specifying units of measure, and labeling axes to clarify the correspondence with quantities in a problem. They calculate accurately and efficiently, and express numerical answers with a degree of precision appropriate for the problem context. In the elementary grades, students give carefully formulated explanations to each other. By the time they reach high school they have learned to examine claims and make explicit use of definitions.

MP7: Look for and make use of structure. Mathematically proficient students look closely to discern a pattern or structure. Young students, for example, might notice that three and seven more is the same amount as seven and three more, or they may sort a collection of shapes according to how many sides the shapes have. They also can step back for an overview and shift perspective. They can see complicated things, such as some algebraic expressions, as single objects or as being composed of several objects.

MP8: Look for and express regularity in repeated reasoning. Mathematically proficient students notice if calculations are repeated, and look both for general methods and for shortcuts. For example, upper elementary students might notice when dividing 25 by 11 that they are repeating the same calculations over and over again, and conclude they have a repeating decimal. As they work to solve a problem, mathematically proficient students maintain oversight of the process, while attending to the details. They continually evaluate the reasonableness of their intermediate results. (Text from Standards for Mathematical Practice, © Copyright 2010 National Governors Association Center for Best Practices and Council of Chief State School Officers. All rights reserved.)

Each of the 5 Principles is associated with several of the Practices, and each strategy within the Principles includes a list of the MPs connected with it. Figure 2.2 at the end of this chapter also helps you cross-reference the Practices with the strategies and Principles.

THE 5 PRINCIPLES OF THE MODERN MATHEMATICS CLASSROOM

The eight Standards of Mathematical Practice describe performances and habits of mind that teachers want students to exhibit. Although they can and should be explicitly taught, if this explicit instruction is done in isolation, the Practices suffer the same fate as the algorithms and definitions we teach: becoming just one more thing that students have to remember, but that they will have difficulty applying and transferring to new situations.

The Practices will thrive, however, if students learn them in an environment designed to support them. Through two decades of teaching and supervising elementary math, I have developed a framework for such an environment, which I call the 5 Principles of the Modern Mathematics Classroom. A classroom culture built on this framework allows your students to grow into mathematical thinkers and sophisticated problem solvers. A modern mathematics classroom incorporates all 5 of these Principles:

Conjecture. In a traditional mathematics classroom, the primary goal is for students to get the right answers to questions and exercises. In a modern mathematics classroom, conjecture is encouraged, students ask most of the questions, and the answer to a question is very often another question. Inquiry is important, as is problem finding.

Communication. In a traditional classroom, communication is primarily one-way: the teacher explaining a procedure or algorithm to students. In a modern mathematics classroom, students must learn to communicate frequently about problems and how they solve them. They focus on vocabulary, writing, and metacognition. The essence of mathematics communication is the formulation and support of mathematical arguments.

Collaboration. In a traditional classroom, students work alone, and the emphasis is on an individual's skill fluency. Modern mathematics classrooms are all about the "we." Group work is far more prevalent than individual work, and students are encouraged to share ideas, answers, and to ask for help. Although there is a time for individual performance, in a problem-solving culture, the other students are cheerleaders instead of competitors.

Chaos. In a traditional classroom, neatness and order rule the day. Students must learn a procedure and then replicate it with mechanical precision. In the modern mathematics classroom, real problems require experimentation, false starts, mistakes, and corrections—sometimes over and over again. Although the term "Chaos" may sound sketchy, it simply encapsulates the idea that real math work is messy.

Celebration. In a traditional classroom, recognition is given for right answers and high grades. In a modern mathematics classroom, anything that leads toward a solution is celebrated: finding one small step of a complicated problem, thinking of an innovative approach, even if it doesn't pan out, or making a spectacular mistake and asking for help. Effort is rewarded over achievement, reflecting Carol Dweck's (2006) research on growth and fixed mindsets.

Figure 2.2 helps you cross-reference the Standards of Mathematical Practice with the strategies and 5 Principles outlined in this book. The list describes each chapter of the book and the Standards of Mathematical Practice (MP) each one supports.

FOR PLC AND STUDY GROUPS

1. Teachers have difficult, time-consuming jobs, and it is understandable that they like quick and easy, plug-and-play ideas for the classroom. What are your personal beliefs about the importance of doing the difficult and deep work of developing the underlying classroom culture? What are you already doing to support innovative thinking in your classroom? What obstacles are getting in the way of developing your classroom culture further? Besides this book, what other supports do you need to make it happen?

2. Review again the section on Depth and Rigor. How does Webb's Depth of Knowledge help you think differently about classroom tasks and assessments?

3. Keep a list or collection of every task you ask students to do in a day (or in one subject for a week), including classwork, homework, and projects. Sort the tasks into categories according to the four DOK levels. It is helpful to do this with a group of colleagues. Analyze your groupings. What patterns do you see? Is there a reasonable distribution of tasks across the four levels?

4. What did you discover about your practice by analyzing the tasks? What did you find that reinforced your expectations? What surprised you? What will you do differently as a result of this analysis?

5. Take any Level 1 or Level 2 task from your analysis and rewrite it to be a Level 3 task. What was difficult about this? How does this help you think about your lesson and task design for future instruction? What can you do to improve your classroom assessments in light of the DOK levels?

Figure 2.2 All Strategies by Chapter

Problem-Solving Culture: Strategies and Practices	Chapter	MP1	MP2	MP3	MP4	MP5	MP6	MP7	MP8
Always Ask Why	3			X					
Check, Recheck, and Double-Check	3	X					X		
"I Guess" Is a Declaration, Not an Abdication	3			X				X	X
Metacognitive Questions	3		X	X					
Never End With the Answer	3			X			X		X
Pattern Watch	3				X			X	X
Problem-First	3	X		X					
Conversations Are More Important Than Computations	4		X						
Convince Me	4	X		X			X	X	
First and Foremost, Communicate	4	X	X	X			X		X
Say It Another Way	4	X	X		X		X	X	
Use Your Pictures	4		X		X	X	X	X	
Use Your Words	4		X	X	X		X		X
Watch Your Language!	4						X		
Cultivating Hunches	5			X		X			X
Daily Data	5		X			X		X	X
Math: All Day, Every Day	5		X	X	X	X		X	X
Math Workshop	5	X		X	X	X		X	
Model-Eliciting Activities (MEAs)	5	X	X	X	X	X		X	

(Continued)

Figure 2.2 (Continued)

Problem-Solving Culture: Strategies and Practices	Chapter	MP1	MP2	MP3	MP4	MP5	MP6	MP7	MP8
Think - Play - Pair - Share	5	X		X	X				
Think Avengers, Not the Lone Ranger	5	X		X					
Avoiding Filters	6			X					
Changing Scenery With Genius Hour	6	X							
Computer Programming	6		X		X	X	X	X	X
It's Gonna Be Messy	6	X	X			X			
iWonders	6	X	X	X	X				X
The Journey Is More Important Than the Destination	6			X	X				X
No Shortcuts to Shortcuts	6	X	X		X				X
Non-Routine and Unsolved Problems	6	X	X	X	X	X		X	
Teaching Like Video Games	6			X	X	X		X	X
Catch Me If You Can	7		X	X			X		
Celebrate Small Victories	7	X							
Daily Math Edit	7			X			X		X
Do the Unthinkable: Go Gradeless	7		X	X			X		
I Guarantee It	7			X		X	X		
Spiral Feedback	7		X	X			X		X
Validate Effort, Not Answers	7	X		X			X		

3 Conjecture

One major drawback of having students spend their formative years memorizing facts is that facts change.

—Kelly Gallagher,
Veteran Educator and Author

Most kids show up for kindergarten already making routine use of higher-order thought processes. They don't need to be taught how to think. They need to learn how to examine, elaborate, and refine their ways of thinking and put this thinking to deliberate use converting information into knowledge, and knowledge into wisdom.

—Marion Brady, Teacher,
Education Administrator, and Author

THE POWER OF MYSTERIES

I am a fan of television crime dramas, shows like *CSI* and *NCIS*. Each week, I enjoy following along as the investigators gather the evidence and piece together the solution to a puzzling mystery. I try to anticipate where the story is going and to figure things out before the episode concludes. Sometimes, I'm able to work out who committed the crime, but the more satisfying stories contain a surprise twist which changes how I understand the puzzle.

Imagine if one week your favorite crime drama began with a rundown of all the clues, and then explained how they were all connected to the criminal. I know that I'd be pressing delete on the DVR pretty quickly. I mean, what's the point of watching the rest of the show if the mystery is already solved at the beginning?

This scenario is similar to what many students experience in a traditional math class. We, the teachers, lay everything out in front of them, and all they have to do is connect the dots in the way we already prescribed so that they can come up with the predetermined outcome. Only, in this show, we take away the remote, so the students no longer have the option to stop and skip to another show that's more interesting.

THE CURIOUS BRAIN

> Simply put, human brains are curious: we are wired to wonder.

There is a reason that mystery stories are enormously popular and have been since they first appeared in the early 1800s. Simply put, human brains are curious: we are wired to wonder. John Medina, molecular biologist, and author of *Brain Rules*, explains that it comes from our need to explore our environment.

> Babies are born with a deep desire to understand the world around them, and an *incessant curiosity* that compels them to aggressively explore it. This need for explanation is so powerfully stitched into their experience that some scientists describe it as a drive, just as hunger and thirst and sex are drives. (Medina, 2014, p. 247, emphasis mine)

We love mysteries because our brains crave the joy that comes with discovery. Curiosity also improves learning, even for incidental or otherwise boring topics (Gruber, Gelman, & Ranganath, 2014).

For students to have this experience, they must have problems to solve. Though we may think our classrooms offer it, little of what normally takes place in a math classroom has anything to do with problems.

Real problem solvers pursue things they find puzzling instead of waiting for someone to present them with a problem. Psychologists call this *adopting a problem-solving orientation* and *searching through a problem space* (Malouff, n.d.; Newell & Simon, 1972).

Stop for a moment and consider your own experiences in math classes. How much of that experience involved adopting a problem-solving orientation and searching through a problem space? These are both dimensions of the first Principle: Conjecture.

Mathematics teacher David Wees tells a story about searching through his own problem space as a young man. Consider how you might react if a student shared this "wondering" in the middle of your lesson.

When I was 13, I remember discovering a really interesting relationship between numbers. I remember adding up $1 + 2 + 3$ and getting 6, and realizing that this was 2×3. I then added up $1 + 2 + 3 + 4$ and got 10, which was $4 \times 5 \div 2$. Then I had an insight that 6 was also $3 \times 4 \div 2$, and that maybe there was some relationship between multiplying these numbers and getting the sum of the consecutive numbers. After a few minutes of playing around, I confirmed that if I took the last number I added and multiplied it by the next number, and then divided by 2, I always got the sum of all of the numbers.

Later that year I learned the algebraic way of representing this formula; $S = \dfrac{n(n+1)}{2}$. By figuring out the formula by myself, the algebra made much more sense.

Conjecture is both a habit of mind for a student and a cultural orientation within a classroom. It is a focus on questions rather than answers, and it is a sense that there is always more to figure out. In classrooms that honor conjecture, students not only question authority, they do so without fear of repercussions. We frequently hear "I wonder" and "what if," and everyone challenges answers to all but the simplest questions. Conjecture is about options and possibilities, not about the one true path.

 ## Problem-First
MP1 and MP3

Too often in math we lose the opportunity to take advantage of students' natural curiosity when we take a linear approach to instruction. Take a look at any published math text, and find the problems in it. What patterns do you see? Content is likely broken into topic chapters, with each chapter divided into eight to twelve daily lessons. The lessons probably begin with a demonstration of an isolated skill, followed by guided practice of the skill, then a collection of exercises to practice the skill independently. After the exercises, you find a few "application" problems, though in most cases they are really just exercises in disguise. If you're lucky, after all of these exercises you might find a couple of "challenge" problems, usually marked with a star to indicate that they are only for the brave.

If you look at the structure of the chapter, you see a similar pattern. There are a series of skill-based lessons, and then at the end, right before the chapter test review there is one lonely lesson on problem solving. The thing is, it's usually focused around teaching a specific strategy in the same way that the other skills were taught: modeling, guided practice, independent practice with exercises that focus narrowly on the specific strategy.

This format is based on an assumption I've heard many times: problem solving has to come only after all the basic skills are mastered. I have even seen entire course outlines where all of the problem solving is saved for the end of the year, after the state standardized testing is over. Think about what this does to a child's curiosity. What's more intriguing: a collection of skill drills, or a conundrum begging to be decrypted? Why, then, would we leave the intriguing part for last? If a mystery novel were laid out the way we teach math, it would start with all of the collected evidence and end with the crime. How many people would buy that book?

> The cure is simple: put the problem first. Pose a challenge so compelling that students are begging you to help them figure it out.

The cure is simple: put the problem first. Begin your unit of study with the starred problem. Pose a challenge so compelling that students are begging you to help them figure it out. At this point, students have opened the door for you to provide them with the tools and strategies they need. There are neurological reasons to do this. Our brains cannot attend to isolated facts and details. "Normally, if we don't know the gist—the meaning—of information, we are unlikely to pay attention to its details. The brain selects meaning-laden information for further processing and leaves the rest alone" (Medina, 2014, p. 114).

Matt Bramucci did just that. The students in his high school music technology course were neither expert musicians nor problem solvers. In fact, most of them could not read music. One day, as the students arrived for what they thought was going to be an ordinary class, they found Mr. Bramucci standing and staring at a video playing on the classroom television. He appeared to be completely engrossed in the recording, which showed a house decked out in multicolored holiday lights, flashing in time to the Trans-Siberian Orchestra's song "Wizard in Winter."

"What are you doing, Mr. Bramucci?" one of the students asked.

"Shh!" Matt waved his hand impatiently at the student. "I'm trying to figure this out." He continued to watch the video as it played. Students watched with him, and soon the entire class was completely transfixed. With a single whispered sentence, a teacher transformed a simple video into an intriguing challenge: "I wonder if we could do that?"

Students spent the next several weeks researching the technology and figuring out how to connect and program the hardware in order to create a similar light show. Along the way, other questions arose, such as "I wonder if we can make this work with live music, too?" New avenues of exploration opened up, and they learned more about music and the math behind it than they would have if Bramucci had marched them stepwise through the individual skills.

Of course, just planting a problem at the feet of your students is not enough on its own, but by starting with the problem, you provide context, motive, and opportunity for learning the content. To better understand the process that students use when they are solving problems, and to help you teach it better, let's look at the related cognitive processes.

PROBLEM SOLVING IS A CYCLE, NOT A RECIPE

Yale University psychologists Jean Pretz, Adam Naples, and Robert Sternberg (2003) describe the problem-solving process as a cycle (pp. 3–4):

1. Recognize or identify the problem.

2. Define and represent the problem mentally.

3. Develop a solution strategy.

4. Organize his or her knowledge about the problem.

5. Allocate mental and physical resources for solving the problem.

6. Monitor his or her progress toward the goal.

7. Evaluate the solution for accuracy.

In a traditional mathematics classroom, the primary goal is for students to get the right answers to questions and exercises. Almost all of a student's cognitive effort is focused on seeking answers, which takes place during steps 5 and 6. All other steps in the process have already been done for the student. Someone else, generally the teacher or textbook author, has identified and defined the problem, presented an efficient strategy, and organized all of the available information, prior to even presenting it to the student. Even the last stage is ordinarily left to the teacher (or the student with an answer key) to verify the accuracy of the "solution."

Medina (2014) puts this in perspective with an enlightening vignette:

> If you could step back in time to one of the first Western-style universities, say, the University of Bologna, and visit its biology labs, you would laugh out loud. I would join you. By today's standards, biological science in the 11th century was a joke, a mix of astrological influences, religious forces, dead animals, and rude-smelling chemical concoctions. But, if you went down the hall and peered inside Bologna's standard lecture room, you wouldn't feel as if you were in a museum. You would feel at home. There is a lectern for the teacher to hold forth, surrounded by chairs for the students to absorb whatever is being held forth—much like today's classrooms. Could it be time for a change? (pp. 256–257)

While today's mathematics classroom may not be as lecture-centered, it still consists for the most part of teachers presenting an algorithm, students absorbing that algorithm and then reproducing it until the algorithm becomes automatic. Repeat until graduation.

Often in today's mathematics classroom, the focus is on either:

- Drilling of DOK Level 1 (Recall) facts and skills, or
- Rehearsing procedures for more complex problems to reduce them down to Level 1 tasks.

It is like a cooking class where students memorize a collection of recipes and techniques, but never go deeper. They never delve into the principles of flavor and texture, food science, or nutrition, and they never learn the reasons the techniques work.

In a classroom based on the 5 Principles, however, students ask most of the questions, and the answer to a question is very often another question. Students pursue things they are curious about rather than things they are told to solve. The intrinsic joy of discovery, the essence of what it means to learn, is the centerpiece of all activity. In other words, when given a basket of mystery ingredients, students ought to understand enough about cooking to be able to make something tasty and satisfying without a memorized recipe.

> When given a basket of mystery ingredients, students ought to understand enough about cooking to be able to make something tasty and satisfying without a memorized recipe.

This is not to say that the teacher has no role or that students are left to pursue anything and everything that strikes their fancy. The teacher's role in a problem-solving culture is to find or create intriguing problems and to frame them in ways that students find compelling. Let's explore some of the strategies and mind-sets that you can establish to transform your classroom into one that promotes conjecture.

GRADES K–3

This chapter and the next four chapters contain five identified sections labeled like this one, highlighting strategies that are particularly useful in a specific grade band. These sections illustrate what the Principle looks like in those grades. All of the strategies, however, are applicable in any grade. If you are a high school teacher, for example, you can find excellent advice in this section that applies to your students as well. Likewise, the suggestions for upper grades can adapt well to elementary classrooms.

The Principle of Conjecture in the primary grades is all about maintaining and encouraging for as long and as deeply as possible students' natural curiosity about the world.

 ## Never End With the Answer
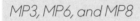*MP3, MP6, and MP8*

A Conjecture-oriented classroom expects students to think and reflect about their reasoning on every problem. Accordingly, with few exceptions no answer to any question should stand on its own in a math lesson. Even if it is correct, the teacher should follow the response with another question. In Grades K–3, I suggest these types of questions:

- Why do you think so?
- How do you know?
- How did you get that answer?
- Why did you solve it that way?
- Are there any other ways to answer it?
- What was hard about solving that problem?
- How did you overcome the difficulty?
- What did you use to help you solve this? How did it help?

At first, students will get annoyed or frustrated at constantly being challenged and asked "why," but it won't be long before the mindset becomes ingrained and students start asking these questions of themselves and each other. Students learn that the "end" of a problem is really just the start of the next phase.

This is also a simple and effective way to add cognitive depth to a textbook exercise. An exchange in a second-grade classroom might follow this example. The students have all had a few minutes to work on this problem:

Lee brought 8 cupcakes to the party. Maria brought 9 cupcakes. Each person at the party ate one cupcake. After the party was over, there were still 3 cupcakes left. How many people were at the party?

Teacher: "OK, so who can answer this question?"

Six hands are raised immediately. The teacher waits a few more seconds while several other hands appear.

Teacher: "Mark?"

Mark: "Twenty?"

Teacher: "You sound unsure. How do you know it's twenty?"

Mark: "Because that's the answer I got when I added."

Teacher: "So why did you solve it that way?"

Mark: "I took all the numbers and added them."

Teacher: "OK, but why does it make sense to add them all?"

Mark: "I don't know."

Teacher: "Can anyone else help explain? Anita?"

Anita: "Well, two different people brought cupcakes and put them together, so you have to add."

Teacher: "OK, tell me more."

Anita: "Uh, but I don't know what to do with the three that are left at the end. I don't think you're supposed to add those."

Teacher: "Why not? Why doesn't that make sense to you?"

Anita: "Well they were left over. Doesn't that mean we should subtract them?"

Teacher: "Mark, what do you think?"

Mark: "Now I'm not sure."

Teacher: "OK, so what seems hard about this to you?"

Mark: "I thought I was supposed to add the ones that were left, but Anita says we should subtract."

Teacher: "So, it sounds like you're trying to figure out what operation to use with the numbers in the problem. Is there another way you could figure out what's going on?"

Mark: "Well, maybe if we had a party with some cupcakes."

Laughter in the classroom.

Teacher
(laughing also):   "That's an interesting idea. Is there a way we could do that?"

Mark
(surprised):   "Really? We can have a party?"

Teacher: "No, we're not going to have a party, but how could we use that idea to help solve the problem?"

Anita: "Maybe we could use pretend cupcakes? Can I be Maria?"

Teacher: "Sure. Mark, would you like to act out the part of Lee?"

Mark: "Yes!"

Notice at no time did the teacher indicate whether any answer or potential solution was right or wrong. He or she did not use leading questions that cue the correct response, either, but very deliberately left the solution path and the answer open to discussion as long as possible.

While this approach helps students increase their ability to reason and discuss math, be aware that it's still very teacher centric. It does allow for increased depth, but there are still only a few students involved in the discussion.

Digital Tools and Resources for Conjecture in K–3

Each grade band in this book includes a brief section describing one or more digital tools that can support the Principle and help you create a problem-solving culture within your classroom. These sections do not give detailed tutorials. I just want to point you in the direction of tools that can strengthen your transformation plan with a few quick ideas for incorporating them into your teaching. Then at the end of the chapter, I highlight one or two tools mentioned in the chapter with a more extended example showing how teachers have integrated them into a 5 Principles classroom.

Part of the goal of the "Never End With the Answer" strategy is to encourage young children to verbalize their reasoning. To extend this idea, try using digital whiteboard apps to allow students to narrate their problem solutions.

One example of this is **Educreations** (http://www.educreations.com). At this site, either using a computer web browser, or on an iPad, teachers and students can use a simple whiteboard to draw and write while recording verbal narration. Even young children who are intimidated by the idea of writing out their math thinking can talk it out by recording. Recordings are private, and only the teacher can access them.

Other similar products are **ShowMe** (http://www.showme.com), which is exclusively available on iPad, and **PixiClip** (http://www.pixiclip.com), which is a web-only application. If you want more flexibility, try using screen casting software, such as **Screencast.com** or **Screencast-O-Matic.com**. Students can use these tools to record everything they are doing on their computer screens as well as voiceover activity. These products probably require more assistance from you to make them work, but they allow the student to use any software on the computer and aren't limited to the specific tools embedded in the whiteboard apps.

Students can share their recordings, and you can use student made videos to prompt inquiry by showing the beginning of a new problem, allowing the class to work on solving it, then showing the student's solution.

GRADES 2–5

In Grades 2–5, the focus shifts from simply nurturing inquiry to guiding and organizing it. Students need help to turn raw curiosity into productive thinking, and the strategies here are designed to achieve that while still encouraging students to continue wondering freely.

 ### Always Ask Why
MP3

As teachers, we are accustomed to asking questions when we already know the answers, and to keep calling on students until we get the answer we're expecting. At that point, we move on to the next question with no further thought.

To create a culture of inquiry and the habits of mind required by the Common Core, expect students to provide explanations of their reasoning. Ask the students their thoughts on it. Ask how they know the respondent is correct. You can also turn some of the responsibility for probing and asking follow-up questions to the students.

> Keep students thinking and questioning. This is even more powerful when it becomes the default process for all student responses.

Keep students thinking and questioning, and encourage reasoning to move far beyond merely guessing. This is even more powerful when it becomes the default process for all student responses. You will notice a predictable pattern when you shift to this strategy. At first, students are thrown off, and many stop volunteering as much, since they are being held accountable for their thinking rather than being able to throw an answer out into the wild and hope for the best. Don't panic; stick with it. Eventually, students gain confidence in their ability to answer, especially if you give them time to think and to share in pairs or small groups before (or instead of) large group responses.

Asking students *why* is the first step away from pure guesswork into the world of metacognition. In Singapore, often held up as the gold standard of mathematics education, metacognition is one key to excellent learning in math. Dr. Wong Khoon Yoong, a mathematician and educator from that country, discusses the importance of metacognition to problem solving:

The Wild Goose Chase in Problem Solving

When solving standard mathematics problems, students normally recall and apply learned procedures in a straightforward way. However, if the problem is unfamiliar, some students simply pick a method and keep persistently on the same track for a long time without getting anywhere. Schoenfeld (1987) described this behaviour as chasing the wild mathematical goose. A different behaviour is also observed: some students jump from one rule to another in a haphazard way hoping to find the correct answer, become agitated, frustrated and finally give up. Teachers who observe both types of unsuccessful behaviours may think that the students have not mastered the skills, and then proceed to re-teach the skills. . . . A more "metacognitive" teacher believes that the students' difficulty may not be with the skills, rather it might indicate a lacking in self-regulation of the problem-solving process. (Wong, 2002, pp. 1–2)

Metacognitive Questions

Lisa Chesser (2014) at the TeachThought blog created a comprehensive list of questions that promote metacognition. Here, I've adapted a few that are particularly helpful for the mathematics classroom in Grades 2–5. Note how these questions promote deeper reflection than those in the last section:

1. What about this problem feels familiar? Why?

2. Why do you think this works? Does it always work? Why do you think so?

3. What about the strategy is working for you? What isn't working? Why?

4. Are there any other similar answers you can think of with alternative routes?

5. What patterns might lead you to an alternative answer?

6. Does anyone in this class want to add something to the solution?

Digital Tools and Resources for Conjecture in 2–5

In Grades 2–5, introduce problems that require lots of messy inquiry in order to help students break away from reliance on right answers and someone else checking their work. In these grades, students also have the perception that all problems have already been solved, and their job is simply to find (or guess) the answer that someone else already got; as a result, many kids ask why they should bother solving this if someone else already did it.

This is a great time to introduce the idea that there are mathematical problems out there which no one, not even professional mathematicians, has solved yet. While most of these are fairly obscure or esoteric, the outstanding website **MathPickle** (http://www.mathpickle.com) translates these unsolved problems into kid-friendly, scaled down versions. Head to the "Unsolved Problems for K–12" section (http://mathpickle.com/unsolved-k-12) to find video and downloadable resources for trying these problems in your classroom.

GRADES 4–7

Moving into the intermediate grades, students are ready for deeper analysis and more complex problems. Though the strategies in this section are useful in the primary grades and I encourage teachers of those grades to think about ways to adapt them, they pay significant dividends in student learning in upper elementary and middle school.

 ## Pattern Watch
MP4, MP7, and MP8

The human brain is a natural pattern-finding machine. It is so good at patterns that we often create them where none really exist. It's why we see constellations in the stars, and why tragedies *always* occur in threes.

One of the characteristics of good problem solvers is that they readily recognize and understand patterns, and that they distinguish real patterns from the ones our brains artificially impose. Often in math instruction, pattern seeking is something that is only done in contrived situations. But, students need to be able to find patterns anywhere, and to detect them when they are irregular. So, begin training yourself and your students to seek and identify patterns in unusual places. Point out patterns in other content areas (current events, creative writing, and weather) and make explicit connections to mathematical ideas. Use patterns to make predictions and use math language to explain them. Blur the lines between content areas.

> The human brain is so good at patterns that we often create them where none really exist. It's why we see constellations in the stars, and why tragedies *always* occur in threes.

Another source of patterns are tessellations, geometric designs which are also known as "tiling the plane." The simplest tessellation is graph paper: squares lined up in rows and columns. But, one can make much more complex and sophisticated patterns of tiles. The works of M. C. Escher are possibly the most famous examples of this type of design. Have students analyze the geometry behind the patterns, and use the shapes to derive important geometric concepts and rules.

Figure 3.1   The Versailles Tiling Pattern

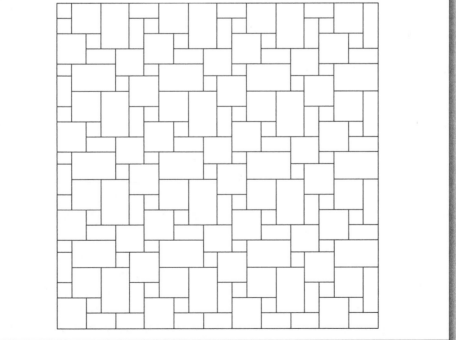

One way to explore these kinds of patterns is to study floor tiles. There are a number of floor tile patterns which are designed to look like random arrangements of shapes, but which on further inspection are actually very simple tessellations.

Consider this common pattern, known as the "Villa" or "Versailles" pattern of tiles (see Figure 3.1 on previous page).

At first glance, the tiles appear to be randomly arranged. But on closer inspection, a pattern emerges, and with some work, students can figure out where the repeats happen. (See if you can sort this out for yourself before reading on. What strategies might help you to visualize the pattern better?)

On deeper analysis, though, one can see the overall pattern is based on a slightly altered square, which is further broken up into smaller rectangles to disguise the larger pattern. Figure 3.2 shows one element of the Versailles pattern.

And Figure 3.3 demonstrates a shaded version to show the original square and how it was altered.

Once you see this shape, it is not difficult to find it in the original pattern and to see that the tiling is essentially just a very large grid of squares.

When using this kind of problem in the classroom it is absolutely essential that you do *not* do what I've just done here: lead students directly to a solution and then provide them with the strategy for solving the problem. Let students explore their own paths, their own strategies, even if they don't lead anywhere fruitful at first. Ask questions that get them to analyze their thinking and whether it is working rather than questions that narrow the field and point them in a productive direction. They learn a great deal from the struggle.

Challenge students to take this concept further: find other shapes that tessellate, such as hexagons or triangles, and create more intricate tiling patterns that appear random. Or research other floor tile arrangements to find the underlying patterns.

Digital Tools and Resources for Conjecture in 4–7

When exploring tiling patterns, it is helpful for students to use manipulatives to enhance their learning. When students have the chance to handle physical objects and move them around to feel and experience the math concepts they are learning, their understanding is deeper and richer. Manipulatives are useful for learning many different mathematical concepts.

You may not always have the right manipulative for the activity you're teaching, however, and Utah State University has developed an excellent, free resource to fill that gap. The **National Library of Virtual Manipulatives**

Figure 3.2 An Element of the Versailles Tile Pattern

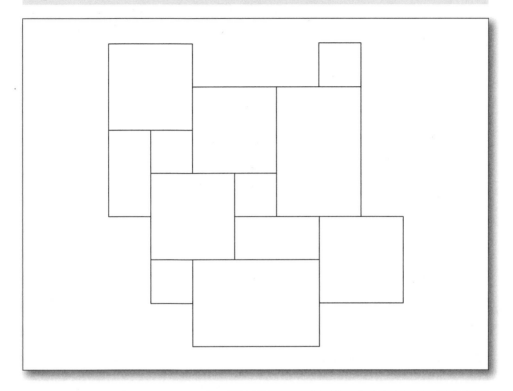

Figure 3.3 The Versailles Pattern Is Based on a Slightly Altered Square

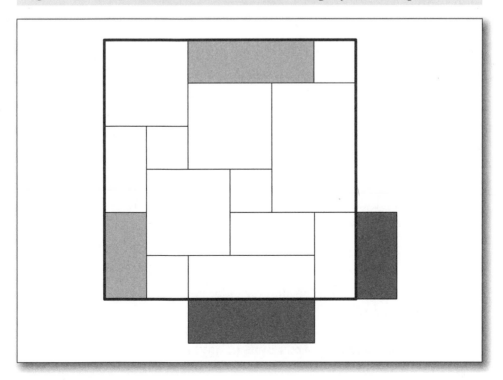

(http://nlvm.usu.edu) contains over 100 different interactive, web-based applets (mini-programs) organized by strand and grade level.

Although virtual manipulatives lose the tactile aspect of physical ones, there are a few advantages to recommend their use:

- The software instantly shows relationships between visual and symbolic representations without tedious computations, allowing the student to focus on the meaning of the concepts
- Students can quickly and easily test different ways of solving a problem and get immediate feedback about what works and what doesn't
- Virtual manipulatives can easily simulate things that are difficult to model in real life
- Recording student work is as easy as taking a screenshot
- Clean up is as easy as closing your browser!

GRADES 6-9

By middle school, students should have a firm grasp of problem-solving fundamentals and know how to bring more structure to the process. Also, these are the grades where mathematics becomes heavily abstract and symbolic, so connecting them back to the Conjecture strategies and techniques that served them well in earlier grades is extremely important.

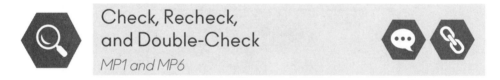

Check, Recheck, and Double-Check
MP1 and MP6

Students need to be trained to understand that when they are done with the computation, they are not yet done with solving the problem. Remind students that they should thoroughly check their work before presenting a solution to a problem. You can introduce this strategy in earlier grades, but middle school students can use it more strategically and deliberately. Try this 3–step approach:

Check

- Did we complete all the steps we planned?
- Is all of our arithmetic correct?
- Did we indicate our answer?
- Did we include the correct units (if any) in our answer?

Recheck

- Did we answer the actual question that was asked? Are we sure we answered the *entire* question?
- Does our answer make sense for that question?
- Can we each explain the work and the thinking that went into the solution?

Double-Check

- Share the solution with another group and discuss. Ask each other questions about the work, anything that is confusing or seems to be out of place.
- Only when everyone is satisfied that the entire solution makes sense and is correct can we say we are done.

Digital Tools and Resources for Conjecture in 6–9

The Check, Recheck, and Double-Check strategy is particularly valuable when used with student-generated problems. The Internet is a deep well of available raw data which students can use to investigate their own questions and form their own problems. One excellent data source is **Data.gov** (http://data.gov), which publishes vast amounts of open, freely available public government data. The American Statistical Society also maintains a list of useful sites for teachers at their website (http://www.amstat.org/education/usefulsitesforteachers.cfm), which includes a section on sources of data. Of course, if you're looking for data about a particular topic, you can often find it at specialty websites. For sports data, for example, you might try ESPN.com or the home page of a particular league. FIFA.com, for example, includes data about how the league calculates team rankings at http://www.fifa.com/world ranking. Students can explore "what if" scenarios for their favorite teams and players.

GRADES 8–12

By high school, students should be routinely asking their own questions and solving their own problems. They should be so used to being asked follow-up questions that they provide the justification before it's requested, and they should be as fluent at making a mathematical argument as they are at adding 3–digit numbers.

Focus in Grades 8–12 on building students' confidence in their ability to ask questions, and using those language and question skills to support deeper understanding of the very abstract concepts they will encounter in algebra and beyond.

"I Guess" Is a Declaration, Not an Abdication

MP3, MP7, and MP8

Ordinarily we don't promote arbitrary guessing in mathematics. We seek precision, accuracy, and efficiency. If there's a shorter, faster way from point A to point B, we show it to students as the "best practice" for mathematical excellence. We value shortcuts, and the shorter the cut, the better.

Yet, for more complex problems, including those we encounter in the real world, the solution path is not immediately obvious. Even though psychologists define problem solving as a cycle, in practice it is more disordered, and problem solvers in the midst of a tough challenge may shift back and forth between steps as they make adjustments, explore possibilities, and reach dead ends.

I recently observed a high school classroom where pairs of students were presented with Duncker's (1945) classic "candle problem." They were given three items: a candle, a book of matches, and a box of thumbtacks, as shown in Figure 3.4. The challenge is to figure out a way to attach the candle to the wall in such a way that when it is lit, the melting wax won't drop onto the floor.

The students at first were stumped. They physically inspected the candle, matches, and tacks, looking for some unusual feature or attachment they might need to use. "Wait, we have to attach this to the wall?" one student asked. "Are we going to light it?"

"No, but you have to attach it in a way that you could light it," the teacher responded.

The students continued to explore the materials, attempting different combinations of candle and tacks, apparently at random, and making little progress.

At this point, the teacher stepped in and showed them the solution to the problem, which is to remove the tacks from the box, tack the box to the wall, then stand the candle inside the box, which forms a drip shield below it, as shown in Figure 3.5.

This teacher's approach is a common one I observe in classrooms, and it is one reason teachers abandon their attempts to infuse more problem solving into their teaching. "I tried doing more problem solving," teachers often tell me, "but the students just couldn't do it. They really need to understand the basics before they will ever be able to solve difficult problems."

Figure 3.4 The Objects Provided in Duncker's (1945) Candle Problem

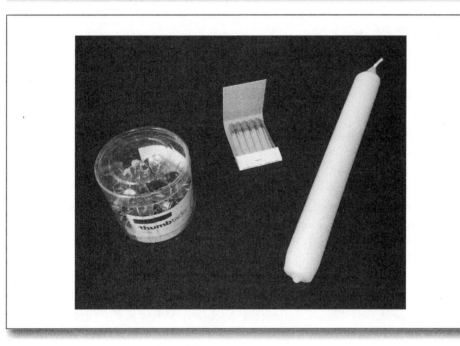

Figure 3.5 The Solution to Duncker's (1945) Candle Problem

> Just as students get better at reading mainly by reading a lot, they will get better at solving problems mainly by solving a lot of problems.

The truth is, however, that it's not a lack of "the basics," it's a lack of experience with problems. This is common in secondary classrooms where students haven't been exposed to a problem-solving culture before. Just as students get better at reading mainly by reading a lot, they will get better at solving problems mainly by solving a lot of problems.

Instead of handing over the answer after one failed attempt, allow students to struggle for a bit. They should be encouraged to make their best attempts at a solution when they do not have a clear or obvious path. Students will not be good at this at first and that's all right.

Of course, you can guide the *process* through carefully selected questions or suggestions. Let's go back and reimagine the candle problem activity from this orientation.

Teacher: *(observing that the students are struggling, and recognizing that they are making random stabs at solutions without fully grasping the problem):* "Let's define the problem. What is the first thing you have to accomplish?"

Student: "Stick the candle to the wall."

Although this is not actually the best first step in the ideal solution, it is a rational one, and the teacher allows them to go with it.

Teacher: "Try making a list of all the ways you might do that. Don't worry about the materials you have for right now, and don't worry about the other requirements of this problem. Just brainstorm ways to attach the candle to the wall."

The teacher then monitors the discussion between the students. It is likely that one of them will eventually suggest something a bit wacky or strange, and his partner will reject it as impossible. This is a good time to step back in.

Teacher: "Don't be too quick to accept or reject any idea. Instead of just throwing an idea out there, why don't you explain in detail why your idea might work. And, instead of just saying an idea is impossible, list all of your objections to the idea. Discuss it more thoroughly."

Let them consider, discuss, and defend their conjectures with each other until they are confident or are hopelessly stuck. If a problem-solving step begins with a guess, the process of testing and arguing their choice will move them toward a solution without you having to short-circuit the learning by injecting the predefined pathway—doing so would be counterproductive, anyway. Learning is an essential part of our survival as humans, and as John Medina puts it, "Our survival did not depend on exposing ourselves to organized, pre-planned packets of information" (2014, p. 253).

Instead of a place where a timid "I guess . . . " is merely the resigned default response when a student is not certain about their answer, let your classroom become one where "I guess!" is the confident declaration of a student who is naming one reasonable way to move forward when the path isn't clear.

Digital Tools and Resources for Conjecture in 8–12

Digital tools can provide structure and organization for student guesses by recording their work or giving immediate feedback on the solution attempts.

For middle- and high-school students, the program **GeoGebra** (http://www.geogebra.org) provides a platform for them to explore and experiment with different geometric and algebraic constructions, testing ideas and visually experiencing in real time how changing parameters and variables affect the shapes and graphs related to them. Recommend that students document and share not just their successful solutions, but their guesses and failed attempts in order to extend their explorations back into the classroom. Another digital tool that can provide a platform for these strategies is **WeLearnedIt** (http://welearned.it). WeLearnedIt is a social learning platform that is aimed at teachers and students who are engaged in project-based learning as a classroom methodology. It is a free iPad app that allows students to provide feedback directly to each other as well as getting feedback from teachers. To encourage Conjecture, create an assignment that contains partially or fully solved problems, then ask students to add written, verbal, or video comments that give their explanations of the work. Students can see and hear each others' responses and give feedback to each other.

TECHNOLOGY INTEGRATION FOR CONJECTURE: CASE STUDIES

MathPickle

Jason Kornoely is an elementary teacher in Grand Rapids, Michigan.

I was teaching fourth grade three years ago when I first used a MathPickle puzzle. There was a broad range of math abilities in my class. I had some very low achieving and low interest students, and

I had very high achieving and high interest students and everything between. I was searching for more creative and interesting ways to bring the beauty of math to my students. I then found MathPickle. It was love at first sight. I spent hours looking through the activities and puzzles. I eventually decided on one puzzle to try with my class to see how it would go over. I chose the Icarus and Daedalus puzzle based on the Collatz conjecture. (See the following puzzle in Figure 3.6.)

I began the day before by reading the story of Icarus and Daedalus so that they would have some background of the story before we explored the puzzle. The next day began with a viewing of a wonderful stop-action animation Lego video on Icarus and Daedalus. This set the sense of urgency to try to save Daedalus. The reasoning behind starting with the Daedalus puzzle was to have students find success relatively quickly; which the majority of student pairs were eventually able to do within fifteen minutes.

Figure 3.6   Icarus and Daedalus Puzzle

The story: Daedalus was a master craftsman in Crete who built the famous labyrinth in which King Minos kept the horrible beast, the Minotaur. Every nine years, the Minotaur demanded human sacrifices, and King Minos sent seven people into the labyrinth to die. One year, Theseus was among those sacrificed. The King's daughter, Ariadne, happened to be in love with Theseus, so she asked Daedalus to save him. Daedalus agreed, giving Theseus a ball of string with which to find his way out of the maze.

King Minos was angered, and imprisoned Daedalus and his son, Icarus, in a tower next to a high cliff. Being the genius inventor he was, Daedalus crafted wings of feathers and wax for he and his son to escape from Crete.

The night before they were to escape, each of the prisoners had a dream. They dreamt that they wrote a number on a rock and threw it off the high cliff. If the rock hit the bottom, they would not escape. If it did not, they would be able to fly away.

Would that work? Could they escape? Only math will tell . . .

The problem: Begin with any number. To find out if Daedalus and Icarus could escape, follow these rules with the number written on the rock. If the sequence ever ends in a "1," the rock hits bottom and the prisoner dies. But, each dream was a little different . . .

Daedalus

- If the number is even, divide it in half.
- If the number is odd, triple it, then *subtract* "1."
- Continue with the new result until you know whether the sequence ends in a "1."

Icarus

- If the number is even, divide it in half.
- If the number is odd, triple it, then *add* "1."
- Continue with the new result until you know whether the sequence ends in a "1."

As the groups were working on the puzzles, there was a very pleasant hum of activity throughout the classroom. I saw some student partners making a chart to keep track of their trials. Other groups of students were discussing which number to begin with based on the results they collected from previous attempts. I heard other students making predictions. Students were discussing if starting with an odd number or an even number makes a difference. I was enjoying the rich mathematical discussions happening among my group.

Every single student was engaged with this puzzle; which is a wonderment to behold in any classroom. Students who struggled were actively working on solutions. High achieving students were eager to forge ahead. The engagement created by this puzzle was enough to satisfy me, however, that there was an unexpected phenomenon beginning to surface.

As some of the sequences started to get a bit larger in scope, student pairs were clamoring to ask me to show them an efficient way to multiply a two digit number with a single digit number. Yes, you are reading this correctly, students were begging me to show them how to multiply numbers quickly. I am not exaggerating when I communicate to you that chills ran up my arms at this astounding turning of the tables—students wanting to be taught.

Finding great success with the Daedalus puzzle, I felt it was a good time to let the students know that there is another puzzle almost exactly like the one they were solving and that it was called the Icarus puzzle. After showing them a few examples, I told them that no mathematician in the world has ever ended up with a result other than "1" when they applied this formula. As the students have come to understand that a result of a "1" is equivalent to plummeting into the ocean waters, they understood why it was called the Icarus puzzle.

To put a cap on this story, I had some of my lowest achieving math students electing to stay in from recess to work on the Icarus problem. I had other students attempt to create a spreadsheet formula to calculate $3n + 1$ for them. Students were bringing the puzzle home and working on it with their families.

From that day on, I was hooked on MathPickle puzzles. I set aside our Friday math time to work on problem-solving puzzles. I have been using these puzzles in my classroom ever since.

The benefits of using Dr. Hamilton's Icarus and Daedalus puzzle were clear. Students with widely varied math abilities were engaged.

The lower achieving students were empowered by the fact that if no mathematician has ever solved the puzzle, why not take a shot at it? The higher achieving students enjoyed the challenge, to be sure. The interest level was certainly present. They loved the connection to the myth. I think the literary connection gave them a tangible reason to solve the puzzle. The most impressive result from using this puzzle was that my students wanted to learn a quicker method for multiplying. I share this story with colleagues across my school district and their reactions are always positive.

MathPickle puzzles make teaching math fun. MathPickle puzzles build confidence in my students.

WeLearnedIt

WeLearnedIt has been adopted in hundreds of schools, in all different types of schools, and with all different groups and grades of students. Adam Bellow, former teacher and creator of the app, describes how an educator in North Dakota implemented WeLearnedIt to help support student inquiry and problem solving with students.

In Mrs. Devlin's 9th grade math class, WeLearnedIt serves as the digital hub for students to propose projects, capture work, gather feedback, and create a portfolio of student work to show academic growth over time. In October, Devlin introduced a unit from the Buck Institute for Education website, PBLU.org, called "Mazer Tag" (http://pblu.org/projects/mazer-tag).

Mazer Tag challenges students to create and play a game where they can guide a laser projection through a maze of obstacles by placing mirrors on the board. This year, Devlin used WeLearnedIt to take this project to the next level. The students were split into teams of two and challenged to create a board for their peers to solve. This is a project that the teacher had created once before, but because of the advanced capturing and annotation tools in WeLearnedIt, she was able to better track the thought process and results in how the students used the principles of geometry to solve each of the levels. The finished product was then shared with a class in Virginia and they were able to collaborate on a virtual board and played a live match where the classes worked together on figuring out the solution while directing their counterparts to place mirrors at specific angles while watching on Google Hangouts.

Additionally, Devlin asked them to pick a math concept that they wanted to explore in greater depth and create a game that they could share with the class and other PBL (project-based learning) classes around the country, further applying the principle of Conjecture. Allowing students the choice to pick both the math concept they wanted to explore and then the means by which they would gamify this concept for their peers, breathed new life into the project. Devlin reported that the student games were truly engaging and ranged from variations of Mazer Tag to video games created in Scratch (a creative online learning community) that were guessing games that allowed the students to study and then program probability.

Finally, the students used WeLearnedIt to not just post their final work, but to share their drafts, pose questions to their peers and teachers, and give and get reflections on their work and the work of their peers.

FOR PLC AND STUDY GROUPS

1. Review the seven steps of problem solving as described at the beginning of this chapter. Does your instruction about problem solving more closely resemble a recipe or a cycle? Which of the steps do you typically allow students to do? Which do you most often do for them? What can you change in your instruction to give your students more opportunities to experience the steps that they don't typically do?

2. The physical environment of the classroom is as important as the intellectual one in establishing a culture for problem solving. Think about what you might do with the arrangement and use of space, walls, storage, and decor to set the right tone as soon as students arrive in the room. What physical features do you think would make the biggest difference in promoting a mindset of Conjecture?

3. One natural outgrowth of promoting Conjecture is that students begin to challenge you as an authority. It is important to maintain an open and safe environment for students to ask questions while still maintaining an atmosphere of respect and a focus on academics. How can you be proactive to prepare for the loss of control that comes with student-led inquiry? What procedures can you put into place to keep your classroom positive and productive?

(Continued)

(Continued)

4. Teachers often feel pressure to shift away from inquiry and student-initiated questions toward direct instruction and test preparation. This is especially so in districts where teacher evaluations are based on student test scores. How is a culture of Conjecture compatible with student performance on state tests and other measures of accountability? What steps can you take to promote Conjecture while still ensuring that your students meet state math standards? How can you communicate to parents and administrators that your approach can accomplish both?

5. Most teachers discourage guessing. What aspects of that strategy concern you? What parts of this approach do you think work best for your style of instruction? How can you encourage appropriate guessing while helping students to hone in on solutions to problems?

4 Communication

To effectively communicate, we must realize that we are all different in the way we perceive the world and use this understanding as a guide to our communication with others.

—Tony Robbins, Motivational
Speaker and Life Coach

If you can't explain it simply, you don't understand it well enough.

—Albert Einstein, Theoretical Physicist,
Authority on Philosophy of Science

THE MYSTERIOUS MATHEMAGICAL MIND TRICK

"Go right to your seats and get ready for math," I said as I brought my fifth graders back to the classroom from lunch.

"What's that, Mr. Aungst?" asked Kelly, pointing at a large manila envelope propped on the chalk tray. Several students paused to look. Momentarily confused, I turned and saw a mysterious and somewhat ominous directive written in large, red, block letters: "DO NOT OPEN YET."

I scratched my head. "I'm not sure. I didn't notice that before. It wasn't there when I left to pick you up from the cafeteria. Very odd." I started to move toward it, then hesitated. "No, I suppose we should follow the directions and not open it yet."

"When will we open it?" Kelly asked.

"I imagine we'll know when it's the right time."

A few minutes later, everyone had settled into their seats. "You know, this mysterious envelope has me thinking," I said. "Who likes magic tricks?" Lots of hands went up. "Great! How many of you believe that I can read minds?"

There were a few chuckles around the room, and a couple of skeptical looks, but no hands this time.

"I really can. It's a necessary skill for teachers. Would you like me to prove it?"

I could sense a few students were intrigued. A few said, "Yes!"

"OK, then. Everyone get out a clean sheet of paper and a pencil. I'm going to give you a series of instructions. These steps are designed to clear your thoughts so that it will be easier for me to read your mind." I began to give them instructions. In order for you to experience it from the student's perspective, follow along with this adaptation of the classroom activity. At each step, start with the answer from the previous row. Instead of using a classroom math textbook for the last step, use this book:

First, write down any two-digit positive integer.	
Multiply the number from the first row by nine.	
Add the digits of the product. If the sum has two digits, add them again. Call this sum "W."	W:
Multiply W by 13. Call this product "P."	P:
Turn to page number P in *5 Principles of the Modern Mathematics Classroom*. Ignoring any headings, begin on the first line of the main text on that page and count to word W. Write this word here.	

When doing this lesson with students, I'd typically have them use their math text and choose any picture from page P. Let's return to the classroom.

"Has everybody chosen their picture? Great! Now, focus on that picture so I can read your minds." I proceeded to perform a ridiculously elaborate ritual, telepathically gathering all of their pictures, selecting one, and imprinting it from across the room onto whatever was inside the sealed envelope.

Throughout the activity, I had not touched or even approached the envelope. "Kelly, you found the envelope. Can you go open it for us?"

Before I reveal the end of the story, I want you to return to the word you found in this book. Now turn to the next page to Figure 4.1. There you will find a group of words arranged in a grid. Your word is the one at the exact center of the grid, the only one in bold. Am I right? Did I read your mind?

Returning to our classroom, Kelly opened the sealed envelope, revealing a single sheet of paper with a picture on it. "How many of you were thinking of this picture?" Every hand was raised. There were a few shocked faces, but most of the class was now curious.

"See, I told you I could read minds. How many of you believe me now?" Kelly was not raising her hand. "Why don't you believe me, Kelly?"

"How could everyone get the same picture?"

"That's a really good question. Maybe you all started with the same number?" A quick check proved that theory wrong. And yet everyone ended up on the same page. Just like you located the ninth word on page 117 of the book in your hand.

So did I really read their minds? And yours? Of course not. I did say at the beginning it was a trick.

"And that's my point," I told my students. "Teachers cannot read minds. And yet so much of our job involves understanding what you are thinking. So, if I cannot read your mind, what's the only possible way for me to know what you're thinking?"

"If we tell you?" Kelly responded.

That is why Communication is the second Principle of the modern mathematics classroom.

FIRST AND FOREMOST, COMMUNICATE

We have established that math is not primarily about computation or a checklist of content and skills. It is about solving problems, and problems are solved through communication.

In a classroom based on the 5 Principles, students must learn to communicate frequently about problems and how they solve them. They focus on vocabulary, writing, and the formulation and support of mathematical arguments.

Therefore, a math classroom should first and foremost be a communication classroom, and should be filled mostly with talking and writing. Most math learning takes place when students are talking to each other and writing—in English, not just math symbols. Students should explain, argue, defend, critique, and discuss ideas.

> Most math learning takes place when students are talking to each other and writing—in English, not just math symbols.

Figure 4.1 Words for Use in the Mysterious Mathemagical Mind Trick

swift	culture	some
the	**also**	below
math	guess	program

Students should write on a daily basis, produce longer math pieces weekly, and share their writing with multiple audiences.

We are going to borrow a lot in this chapter from language arts instruction, including best practices for English Language Learners, because teaching communication in math is very similar to teaching communication in English. The context and content may be different, but the goals are the same: to successfully convey ideas between two people.

SIX LEVELS OF MASTERY

Research supports the premise that communicating about math is essential to full understanding. David Sousa (2008) describes six levels of mastery that a student must move through in order to learn and retain mathematical concepts (see Figure 4.2 for an example illustrating the six levels). Pay attention to how much of math mastery is dependent on communication:

- Level One: Connects new knowledge to existing knowledge and experiences
- Level Two: Searches for concrete material to construct a model or show a manifestation of the concept

Figure 4.2 Illustration of Sousa's Six Levels of Mastery

Level One: The student recognizes that fractions are related to division, and that $\frac{3}{4}$ is the same as $3 \div 4$.

Level Two: The student measures out three cups of sand into a container, then divides the sand into four equal amounts and measures each pile to see that it contains $\frac{3}{4}$ of a cup.

Level Three: The student draws a picture showing that $3.00 can be divided into four equal amounts by exchanging dollars for quarters and rearranging them.

Level Four: She writes $\frac{3}{4} = 3 \div 4$ and $3 \div 4 = 0.75

Level Five: The student then solves the following problem, "Elana's band teacher is holding a three-hour dress rehearsal right before the winter concert. The band has four songs to practice, and the teacher wants to spend the same amount of time on each one. How long will the band practice each song?"

Level Six: The student explains orally how fractions and division are related to a peer or to the class, or explains the reasoning behind her solution to the band practice problem.

- Level Three: Illustrates the concept by drawing a diagram to connect the concrete example to a symbolic picture or representation
- Level Four: Translates the concept into mathematical notation using number symbols, operational signs, formulas, and equations
- Level Five: Applies the concept correctly to real-world situations, projects, and story problems
- Level Six: Can teach the concept successfully to others, or can communicate it on a test

(Sousa, 2008, pp. 169–170)

GRADES K–3

 ## Watch Your Language!
MP6

Several years ago, I had the opportunity to visit Madrid. It was the first time I had been in a foreign country where the local residents didn't speak any English. My own knowledge of Spanish is limited to counting to ten, saying thanks and good day, and a few colors and food items. (Thank you, *Sesame Street*.)

It was an extremely enlightening experience. I knew for the first time what it was like to need to communicate, but not be able to express myself. It was frustrating for me, and for many of the *madrileños* with whom I interacted.

For young children, learning math is like learning a second language. Complicating things, we use English words for much of the work, but we also mix in an entirely different symbol set with meanings of its own. As adults, we fluently read the symbolic language of basic math and translate it automatically into its English equivalents. When we speak math aloud, we use English words. It's easy to forget that for our students, they are still two different languages.

> As adults, we fluently read the symbolic language of basic math and translate it automatically into its English equivalents. It's easy to forget that for our students, they are still two different languages.

Our colloquial use of some words is also different than their more precise use in math. Further complicating things, math has specific academic language for particular situations that we speak about more broadly in common conversation. Let's take the word "answer." Beyond its meaning as the response to a question, we also use it to mean the result of a mathematical computation, or the solution to a problem. However, mathematics has more precise terms that students need to begin using as soon as they begin learning math. Use the word *solution* when talking about problems. For computations, teach the terms *sum, difference, product*, and *quotient*, and help students understand and ingrain their meanings.

Teach academic math vocabulary the same way that you would in a language arts class. To accomplish this, Marzano and Pickering (2005) provide a straightforward six step process:

1. Explain the new word. This must go beyond reciting or copying a definition. Use examples, imagery, and nonlinguistic representations.

2. Students now restate in their own words. Repeat this word both verbally and in writing, as appropriate for the developmental level of the child.

3. Students should now create their own nonlinguistic representation. This can be a picture, a model, a demonstration, or their own symbolic representation. Some students with lots of background knowledge may give you the standard symbols; acknowledge this, then encourage them to come up with their own representation so that they can show what they know about the word.

4. Regularly engage students in activities that allow them to deepen their understanding of the words, such as categorizing, sorting, semantic mapping, analyzing root words and affixes, or creating analogies and metaphors.

5. Students should also have discussions about the words with partners and small groups. They can explain their understanding, their visual representations, or new ideas they have learned about the meaning of the word. This can and should happen by itself and also in connection with the activities in Steps 4 and 6.

6. Play games to explore and review the vocabulary terms. This could also happen as part of games played to review and understand other math concepts, but vocabulary should also receive its own dedicated review time.

We also need to explicitly teach the symbolics of math as its own language for communicating mathematical ideas in short, simple ways. Talk about how you are really translating between two languages when you discuss something in English and then show it in mathematical symbols.

This became clear to me several years ago during a lesson on Egyptian hieroglyphs with my fourth grade gifted class. The activity took place immediately after I taught a fifth grade math lesson on equations with variables. While explaining to my fourth graders that the Egyptians often wrote hieroglyphs out of order, and that some of the symbols represented sounds, some ideas, some were modifiers, and some had different meanings depending on the context, I realized that this is exactly how our mathematical symbol system works.

One of the frustrations that I have when teaching math is that students tend to read from left to right, and often when they get to something that hangs them up, they just stop there and don't try to figure it out. Beyond the most elementary number sentences (2 + 3 = 5), this approach doesn't work well. It is essential for students to learn that sometimes you read from right to left, sometimes you read from the middle out, and sometimes you have to piece different parts together until the whole equation makes sense.

It's routine in elementary math instruction to ask students to write numbers in multiple forms so that they can see the relationships between them. For example:

Standard form: 728

Expanded form, explicitly representing the place values: 700 + 20 + 8

Word form, showing how we speak and read the number: seven hundred twenty-eight.

But, it's rare that we do the same for any of the larger symbolic representations of math, like expressions and equations. Make time in your math instruction to try something like this:

Standard form: 11 + 24 = 35

Word forms:

- eleven plus twenty-four is equal to thirty-five
- the sum of eleven and twenty-four is thirty-five
- thirty-five is equal to the sum of eleven and twenty-four
- when you combine the quantities eleven and twenty-four, the total is thirty-five

See how there are several ways to express the same symbolic statement in English, and the English statements don't necessarily correspond to the symbols in left-to-right order? Students need to see that all of the word forms are versions of the same

> Ask students to practice translating their words into mathematical symbols and vice versa.

mathematical concept, expressed by one symbolic statement. Ask students to practice translating their words into mathematical symbols and vice versa.

Digital Tools and Resources for Communication in K–3

Although it is important to get kids writing about math as early and often as possible, the focus during these years should be on building verbal fluency with mathematics. This emphasis leads to much better written work in the upper elementary grades.

Any means of digitally recording students is a terrific way to help them improve these skills. Recordings can be used for math instruction in a number of ways:

- Make a video of students talking about their math work, then have them watch their own explanations to reflect on what makes sense and how they can explain things more clearly.
- Keep an archive of the best examples of student math talk and use them as demonstrations in addition to your own modeling and think aloud strategies.
- Create a class podcast or video blog to share student math work with parents. This creates a different audience for the communication, since parents weren't in the room during the lesson and will need a different kind of explanation to understand what is going on. This also opens up the possibility of two-way communication about math.
- Have students record their own instructional videos about vocabulary for either younger students just learning the terms or as review for students in the same grade.

GRADES 2–5

 ### Conversations Are More Important Than Computations

MP2

In the middle of a demonstration activity I was doing in a sixth grade classroom, the classroom teacher pulled me aside. "I see what you're doing here and I like the idea," she said. The students were working in small groups to solve a small, but challenging problem: *with a small box and a large sheet of paper, figure out how to completely cover the box with one piece of paper so that there are no gaps or overlaps.* There was a lot of activity, a lot of mess, and most importantly, a lot of student talk. Her students were talking animatedly about the problem. "I understand this," she continued, "but I get very uncomfortable when students are out of their seats, moving around, and talking."

The math concept behind this activity is understanding that a prism, a three-dimensional figure, can be represented by a *net* of two-dimensional shapes. For example, a cube can be represented by the net in Figure 4.3.

Imagine folding the shape along each line. The resulting figure is a cube. Many people have difficulty visualizing this result without actually cutting it out and folding it, so I recommend you go through the process. You can also view an animation of it on YouTube at http://youtu.be/4mvJEn5Kst4.

A cube can also be represented by either of the nets shown in Figure 4.4. Try cutting these out and folding them to see how they become a cube.

Figure 4.3 A Net Representing a Cube

Figure 4.4 Two More Nets Which Form Cubes When Folded

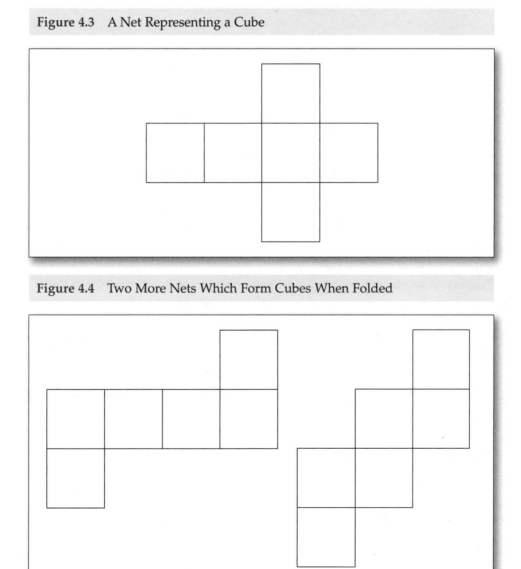

Students have difficulty mentally manipulating these shapes unless they have extensive experience handling physical models and folding the nets. Asking the group to start with the box and determine how to generate a shape that can wrap it gives them experience with the actual manipulation which in turn helps them when they deal with the abstract nets later. Moreover, the students are also building a common language for expressing ideas about these shapes and about problem solving in general.

> The conversation they're having is where the learning takes place. . . . It's the thinking they do while they're talking that leads to an understanding of the math.

I tried to reassure the anxious teacher. "The conversation they're having is where the learning takes place. Listen to what they're saying to each other. They are discussing options, debating what may or may not work, encouraging and helping and challenging each other. They're even arguing. It may seem disorganized, but it's the thinking they do while they're talking that leads to an understanding of the math."

"But, I feel so much pressure to cover all the content before the state test," she said, echoing what I hear from a lot of teachers when I talk to them about the 5 Principles.

I told her that I also felt the same pressure as a teacher. There's so much content to cram in before THE TEST that it's extremely tempting to take shortcuts and bypass the development of understanding. In this lesson, I spent an entire hour letting the students explore and discuss their solutions to this one problem. It took me another four hours of follow-up lessons to build enough experience and understanding with the boxes and paper and similar activities that they deeply understood nets and could fluently move from the concrete to the abstract. The students' conversations were crucial to that understanding and fluency.

"We know that task-related talking is important for learning the vocabulary of mathematics. Providing students the opportunity to communicate their actions can clarify mathematical terms and phrases" (Sousa, 2008, p. 90). During that first lesson, students struggled to communicate some ideas. As I circulated, I provided some terminology to help. We also closed the first lesson with a discussion about the paper shapes they created. "There is a name for this shape. Mathematicians call it a *net*." Students then wrote their own definitions based on their experience that day. Though these were far from complete, they formed the basis of our conversation over the next several days as we refined and clarified those definitions.

As you encourage these mathematical conversations in your classroom, keep these tips in mind.

1. **Give students abundant opportunities for casual math talk.** Conversation doesn't have to be structured to be productive. Johnson (2010) shares a case study of a molecular biology lab. Although most people would expect the major scientific breakthroughs to occur through lab experiments, "most important ideas emerged during regular lab meetings, where a dozen or so researchers would gather and informally present and discuss their latest work" (Johnson, 2010, p. 61).

2. **Guide conversations through targeted questions and modeling.** Avoid whole group instruction about how to talk about math. Instead, monitor student interactions, joining their discussions and showing them how you think and talk about math. Share insights, use appropriate math vocabulary, and ask specific questions designed to elicit thoughtful conversation.

3. **Provide students with conversation guides to help them when the discussion runs dry.** Figure 4.5 has a list of some good questions and discussion prompts to help you develop a conversation guide for your own classroom. These could be general guides for students to keep handy at all times, but for larger more complex problems, I recommend you develop a specific guide for the problem, just as if you were teaching a novel.

Figure 4.5 Sample Conversation Guide

Below are some questions and discussion prompts you might use to build a math conversation guide for your classroom:

1. What is easy about this problem? Why do you feel it is easy? Do you agree with your partners?

2. What do you think is confusing or complicated about this problem? Why is it confusing or complicated? Do you agree with your partners?

3. What happens if you try the opposite of what you're doing now? It may not work, but it could give you some interesting ideas about the problem.

4. If you gave this problem to an expert mathematician, what do you think he or she might try next? Why do you think that? What happens if you try the same thing?

5. Do you think you are missing some important information or knowledge? What else do you need to learn in order to make progress? Where (or from whom) do you think you could learn it?

6. Tell your partner your craziest idea for what to do next. Or, tell your partner why you think his or her crazy idea might work. Then try it. Talk about what you notice.

7. When you disagree with your partner, try switching sides. List as many reasons as possible why his or her idea could work and yours might not. Then listen while your partner tells you all the reasons why your idea could work. Talk about what you heard and learned.

Digital Tools and Resources for Communication in 2–5

While conversations about math in the classroom tend to be mostly verbal, students in Grades 2–5 can begin using online modes of discussion. Here are two tools that support online conversations about math.

Edmodo. Edmodo (http://www.edmodo.com) is a safe, closed environment where teachers can post questions and topics and students can respond online. Only those with an invitation code approved by the teacher can participate, so the teacher has complete control over participants in the discussion. The site also allows for a parent code so that you can invite parents to observe or even join the conversation.

Google Classroom. Google Classroom (http://classroom.google.com) is a newer entry into the online classroom space. It has fewer features than Edmodo, which can make it simpler for younger students to use and understand. Teachers can post announcements and assignments, and students can submit them online using their Google Drive. Teachers can also allow students to comment and create posts of their own, fostering discussion.

GRADES 4–7

Moving into the middle grades, students are ready to develop strong argumentation skills. A particularly valuable strategy is to ask students to convince you that their solution to a problem is a good one. But, don't be too easy to convince.

Convince Me
MP3, MP6, and MP7

Mason, Burton, and Stacey (2010) present a 3–stage approach to building a convincing argument and exercising students' skills in communicating clearly and effectively about mathematical ideas. I recommend building up to the third stage over time. Allow students to practice Stage 1 for a while before you introduce Stage 2. Once they are comfortable with Stage 2, then you can introduce the more challenging Stage 3.

Stage 1: Convince Yourself. Begin by having students write down a few ideas about why they believe their solution to a problem is good. Encourage them to focus on their reasons for selecting each step, and how they know it works.

Example: Consider the following problem:

A frog jumped on several stones on his way to the pond. He did not land on the same stone twice. The product of all of the stones that he hopped on was 19,635. On which stones did he jump? (Holtz & Malen, n.d.)

Figure 4.6 Frog Pond Problem

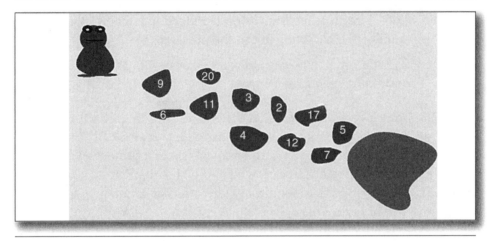

Source: Holtz & Malen (n.d.). Courtesy The Franklin Institute.

Here is how one student responded with her reasons for convincing herself in bold:

- I divided 19,635 by 5, **because the last number is a 5 which means it could be divided by 5.**
- I divided 3,927 by 3, **because if you add all its digits 3 + 9 + 2 + 7 = 21 which is divided by 3.**
- I started dividing by the rest of the numbers until I got the right answer.
- I continued dividing until I got 2 of the numbers that were on the graphic.

Stage 2: Convince a Friend. In Stage 2, students begin to share their ideas with a friend, usually someone they worked with to solve the problem. In this case, their friend is someone who trusts them and their thinking, so the person is not difficult to convince. Most of the work here is on clear communication: expressing ideas in a way that someone else can understand them.

Example: Surina shares her frog pond solution with a classmate, Lamont. He has also worked on this problem, though not together with Surina. The conversation might go as follows:

Lamont: "You said the last number is a 5, but there is also a 7 on the other stone right next to the end."

Surina: "Oh, no, I didn't mean the last stone was a 5, I meant the last number in the answer was a 5. That means I can divide it by 5, so the frog must have landed on the 5 stone."

Lamont: "Oh, I get it now. But, I don't understand where you got the number 3,927. That's not in the problem."

Surina: "That's what you get after you divide 19,635 by 5. I guess I should have put that down."

Stage 2.5: Convince a Mathematical Friend. To take this a bit further and challenge students who master it more easily, try what I call "Stage 2.5." Encourage them to convince a "mathematical friend": someone who trusts them, but really understands math well and wants mathematical explanations for everything. This requires more precise language and more thoughtful use of math ideas.

Example: You listen to the exchange above and encourage Surina to be more precise in her explanation:

Surina: "OK, so what I meant was that the last digit in the final product from the problem was a 5. That means the product is divisible by 5 and the frog must have landed on that stone."

Stage 3: Convince an Enemy. The most sophisticated kind of argument—and the most difficult to construct—is one designed to convince someone skeptical, a person actively looking to pick apart your reasoning and find its flaws.

Initially, the teacher should take this role. Let students know what to expect: I am looking for ways to challenge your thinking. It isn't to make you think you're wrong, but to make sure you have thought about every possibility and have good evidence for what you're saying. Sometimes this makes you go back and revise your plan. That's okay! You will have a better solution and a more convincing argument in the end.

Example: You now step into Surina's and Lamont's conversation as the skeptic:

You: "OK, so I understand your first two steps, but then you just say you started dividing by the rest of the numbers. Did you just pick arbitrary numbers, or did you think about which one to try next?"

Surina: "Well, after I divided 3,927 by 3, I had 1,309. That isn't even, so the frog couldn't have jumped on any even numbers."

You: "How do you know that?"

Surina: "Because as soon as you multiply by an even number, the product will be even."

You: "OK, that makes sense. So why in the last step did you say you had to find 2 numbers that were on the graphic? Is there a specific reason it had to be two?"

And so on. For students to succeed in Stage 3, they need to not only be thorough and precise in their reasoning and language, but they need to anticipate objections and fully understand their math work in order to withstand challenges. Eventually, they will prepare more of their work up front, making sure their reasoning is airtight before they make it public.

Once the students master anticipating the challenges, students can then begin to take on this role for each other, playing devil's advocate against their partners. The ultimate expression of this kind of argumentation is to persuade a skeptic that *someone else's* reasoning is valid. To do this, students need to be able to comprehend and decipher another person's work and thought processes. In our scenario, Surina might give her revised solution to Nadav, who will then have to defend that work to Emma without coming back to Surina first for help.

 ## Use Your Words
MP2, MP3, MP4, MP6, and MP8

Verbal conversations about math are crucial to developing good communication skills. Nevertheless, to take their work to the next level, students need to write about it. With a verbal conversation, there is opportunity for give-and-take, providing feedback to the speaker to clarify statements and fill in gaps in logic. When students write about a solution, they have to consider all relevant information up front, crafting a response that stands on its own.

For more than twenty years, Dr. Annalisa Crannell, mathematics department chair at Franklin and Marshall University, has been asking her college students to write their answers. In her course materials, she includes this explanation:

> Professional mathematicians spend most of their time writing: communicating with colleagues, applying for grants, publishing papers, writing memos and syllabi. . . . It is ironic, but true that most mathematicians spend more time writing than they spend doing math.

But most of all, one of the simplest reasons for writing in a math class is that writing helps you to learn mathematics better. By explaining a difficult concept to other people, you end up explaining it to yourself. (Crannell, 1994, section 2)

> "One of the simplest reasons for writing in a math class is that writing helps you to learn mathematics better."
> (Dr. Annalisa Crannell)

Students must have daily opportunities to write about math. Much of this writing should be focused on directly communicating about problems they are solving and mathematical arguments they are making, but don't limit students to these genres. Nearly any writing form and activity can be used to support and enrich mathematical learning. Here are a few suggestions:

- **Quickwrites.** Ask students at the beginning of a lesson or unit to take one minute to write down everything they think they know about the topic. The goal is to get as much down and fill up as much space as possible. This helps build student fluency with communicating math ideas and using appropriate vocabulary.

- **Individual Whiteboards.** Elementary teachers are very familiar with using individual response whiteboards for allowing all students to complete a math exercise. They promote student engagement and accountability, and allow teachers to make a quick formative assessment. Try using these for brief writing tasks as well, such as defining a key term, describing the next step in a solution, explaining a process, or writing an answer to a teacher question.

- **Math Reporter.** Have students work in pairs to write solutions and explanations to problems. One student takes the role of mathematician, and the other student is the math reporter. The mathematician verbally explains the process and steps in a solution, and the reporter takes notes, ask questions, and collects information. The reporter then writes a news article on the problem.

- **Exit Tickets.** On the way out the door, have students hand you a one-sentence summary of the math they used, an idea they learned, or a vocabulary definition.

- **Math Fiction.** Don't ignore the possibility of combining math and fiction. There are many examples of children's and young adult literature that use mathematical ideas as a launchpad for intriguing stories. Middle grade students enjoy writing books for young children. Why not ask them to teach an interesting math idea through a story?

- **Math Poetry.** Poetry lends itself naturally to math connections. The rhythms and patterns are obvious ways to work math into student writing, but students can also express math ideas in poetry. For example, challenge students to write poems about circles that use the digits in pi as a template for the number of words or syllables in each line.

- **Math Research.** So often we focus entirely on math content in our math classes while ignoring the rich history and fascinating personalities of mathematics. Have students research the origins of our number system, or investigate ways that math was used in events they are studying in social studies. Encourage them to find out about the people who originally invented or proved the math concepts they are learning. There are abundant sources of information about math history and famous mathematicians on the Internet. Pay special attention to women and people of color to give your students diverse, positive mathematical role models.

Digital Tools and Resources for Communication in 4–7

One of the best ways to extend your students' math writing beyond the classroom and provide them with an authentic audience for their work is through blogging. Classroom and student blogs are a terrific platform for students to practice their own writing skills and share their mathematical thoughts with the world. Here are a few tips for blogging with students.

Choose a platform that fits your needs. General platforms like **Blogger** (http://www.blogger.com) and **Wordpress** (http://www.wordpress.com) are flexible and have lots of options. **Kidblog** (http://www.kidblog.org) and **Edublogs** (http://www.edublogs.org) cater directly to the education market with additional security features.

Don't grade blog posts. Blogs should be an informal place to test out thoughts and share partially formed ideas for feedback and discussion. Encourage students to post their work even if they haven't completely worked out the best way to express their thoughts. The real power of a blog post comes from the dialogue generated with the audience through comments.

Share your student blogs with parents and administrators. When your students realize that others besides you can read and respond to their writing, they will put more value on finding the best way to say something and to make clear and relevant arguments. Asking parents and administrators to visit your students' blog and enter the conversation is an organic way to keep them up to date on your classroom activities.

Partner with other classrooms. This can be a very effective way to get students to collaborate and connect. Look for teachers in your school or district who want to work with you. It is even more powerful to connect with classrooms across the country and around the world. Though you can certainly take turns reading and discussing each other's blogs, consider a joint venture where your students work with others to build a team blog.

Integrate math blogging into the ELA and technology curricula. If your classroom is self-contained, this is fairly easy to do. If, however, your students learn English language arts and technology content with other teachers, consider partnering with those teachers to teach writing and digital literacy skills across disciplines. Blogging allows students to learn about writing good titles, tagging and categorizing posts, hyperlinking to credit sources, and using multimedia.

GRADES 6-9

As students get older, communication about math needs to become more technical and precise. Students also need to deepen their understanding of the relationship between the math language of symbols and equations and the English meaning of those symbols.

Use Your Pictures
MP2, MP4, MP5, MP6, and MP7

Images are another tool often used to communicate in mathematical contexts. In particular, infographics are a wonderful way to quickly and concisely communicate mathematical content and relationships. Consider incorporating the study and creation of infographics into your regular math instruction.

Infographics are a genre of communication by themselves, so it's important not to simply throw some data at your students and ask them to make infographics. Instead, build infographic literacy in stages.

1. **Read and interpret existing infographics.** An infographic is essentially any visualization of data. Simple graphs and charts certainly meet this definition, so learning to read and interpret these is an important step. However, don't stop with the infographics built into your math program.

Expand students' horizons by exploring other kinds of infographics. Figure 4.7 gives a list of some websites that have good sources of different infographics.

Figure 4.7   Sources of Infographics

> Be aware that some of the sites listed below are not explicitly created for educators or children, so you may find content that is not suitable for the classroom.
>
> - **Kathy Schrock's guide to infographics**, which includes resources for teaching at http://www.schrockguide.net/infographics-as-an-assessment.html
> - **Daily Infographic** provides high quality infographics, with a new one posted each day at http://dailyinfographic.com
> - **Mashable** frequently posts articles incorporating infographics. These are archived at http://mashable.com/category/infographics
> - **Good Magazine** also has an entire section devoted to infographics on current topics at http://magazine.good.is/infographics
> - **Cool Infographics** at http://www.coolinfographics.com and **Information Is Beautiful** at http://www.informationisbeautiful.net both periodically share interesting visualizations of data.

2. **Analyze infographics more deeply, make inferences about them, and critique them.** Beyond just finding the data in the infographic and understanding its meaning, students should start to analyze both the data itself and the techniques used to communicate it. For example, how does the graphic in Figure 4.8 on page 68 use shapes and shading to communicate ideas about good nutrition and healthy eating? How accurate is the graphic's information? Is there anything misleading or confusing about how the image is composed?

3. **Look for other non-traditional visual means of sharing and understanding data.** For example, examine a photograph of a crowd at a sporting event or political rally. What meaningful data can you extract from this? For example, estimating the size of the crowd, or the proportion of fans of different teams, or counting individual heads and signs. How can the image help you understand the event? Or give students a transit map and see if they can make any inferences about route schedules, population, or geography.

4. **Give students opportunities to experiment with turning data into visuals.** Have them work with data sets available online, data from local sources, or data they collect themselves. Though it is tempting to jump immediately to digital tools, and I have included links to several good ones in the next section, students should work analog first to get a feel for good design. Sketch ideas, or use cut paper and found images to build interesting visualizations.

Figure 4.8 USDA MyPlate Icon Illustrating US Dietary Guidelines

Source: http://choosemyplate.gov

5. **Take infographics into the third dimension.** Some data lends itself better to a 3D model than a flat graphic. Have students build models using blocks, construction toys, or clay. The work of photographer Marion Luttenberger shows some unique ways to create engaging and exciting infographics using real-world objects and even people. You can see examples of his infographic work at his website, http://marion-luttenberger .squarespace.com/#/infographics.

Digital Tools and Resources for Communication in 6–9

Students can create digital infographics using any drawing or graphics tool, and most spreadsheet programs can create basic graphs and charts from data sets. Students can learn a lot by experimenting with how programs like Excel and Numbers translate data into images.

Online digital tools for creating and sharing infographics can take students beyond these simple resources. Here are two of the easiest websites for students and teachers to learn and use. Both sites provide free accounts for teachers and students. The free versions generally limit the templates and tools that are available, although in both cases, even these limited features are adequate for most classroom applications. Additionally, education discounts are available for their paid versions if you find that you frequently use these tools.

Piktochart (http://piktochart.com). This site provides robust tools for creating infographics. It includes many icons and images students can incorporate into their work. Piktochart provides extensive tools to organize and customize the design of your infographic. However, with this power comes some complexity, and Piktochart is probably more appropriate for students with more computer experience in creating images.

Infogr.am (http://infogr.am). This site is less comprehensive than Piktochart in terms of design flexibility, but it makes up for this in simplicity and a few features that Piktochart does not provide. Infogr.am is more dynamic, permitting users to embed videos into infographics, and allowing the creation of interactive graphics where viewers can click to change the visual or to switch data sets. In addition, Infogr.am has built-in maps for illustrating data related to states and countries.

GRADES 8–12

 Say It Another Way
MP1, MP2, MP4, MP6, and MP7

Another classroom strategy that builds a culture of Communication while strengthening metacognition and Conjecture is to ask students to restate answers in other terms, or to explain their solutions differently. This strategy reinforces that there are multiple paths and multiple solutions to many problems. It also encourages deeper reasoning, since students may have to stretch to find new ways to express what a problem or solution means. As Ostroff (2012) puts it, "Explaining your thoughts not only communicates your knowledge, but also increases your knowledge" (p. 152).

This activity can take place in multiple ways. For example, you might ask the class to listen to one student's explanation and restate in another way, or you can require students working in groups on a problem to produce not one, but two or three different explanations.

To teach this skill, give students worked problems, including all of the supporting computations. Ask the students, in groups or individually, to provide a commentary on the computations, explaining the rationale behind them. They can document their commentary in a number of ways: with comments in the margins or sticky notes, with highlighted comments in an electronic document, or by recording a video explanation as they point out the relevant places in the math work.

Alternatively, provide a solution with an intentional error somewhere, preferably in one of the steps to the solution rather than a simple computation mistake. Ask students to do an analysis of the implications of the error: how "fatal" was the mistake, and what consequences would follow if they acted based on the erroneous solution? This helps students begin to see math as a tool for decision making, not just as an academic exercise done in isolation.

For another angle, think of the language of mathematics as an explicit translation opportunity. Alexander Bogomolny, a former mathematics professor, gives a very brief sketch on his website of the development of our modern mathematics notation over the centuries as mathematicians struggled to find more precise ways to communicate mathematical ideas. "To help you appreciate the expressive power of the modern mathematical language," he says, "and as a tribute to those who achieved so much without it, I collected a few samples of (original, but translated) formulation of theorems and their equivalents in modern math language" (Bogomolny, 2014, para. 6).

Here is one of the samples he shares, from *Measurement of a Circle* by Archimedes:

> The area of any circle is equal to a right-angled triangle in which one of the sides about the right angle is equal to the radius, and the other to the circumference, of the circle.

It's instructive to process this and see how we have distilled our modern formula for the area of a circle ($A = \pi r^2$) out of a written explanation from one of the greatest mathematicians who ever lived. When we abandon the words and simply apply the memorized formula, much of the original meaning that the notation was intended to communicate gets lost. Use examples from historical mathematical literature and ask students to translate them into our modern notation. The capability to express a concept in English and in mathematical symbols deepens students' understanding and their ability to apply it in different contexts.

> The capability to express a concept in English and in mathematical symbols deepens students' understanding and their ability to apply it in different contexts.

Taking this further, high school students should extend their use of mathematical language into other domains. In life, we use math all the time to support arguments in non-mathematical areas. Mark Barnes, former English teacher and expert in creating rich learning environments in the classroom, says, "If I were doing a cross-curricular activity with a high school math teacher, I might have students explain the concept using abstract vocabulary—generalizing to make it as simple as possible. Then, I'd have them add details that bring the abstract language into focus." He gives this example—first present the abstract hook: *School uniforms will eliminate violence and theft at our school.* Have students discuss the topic, including any mathematical concepts that are involved in constructing the argument. Then, bring focus by adding the mathematical details: *This study says that there are eighty percent fewer fights in schools with uniforms.* The argument needs to include a mathematical analysis of the study and the data reported.

Ultimately our goal should be for students to be as fluent in the academic and technical language of math as they are in everyday English. This can only happen if they are immersed in an environment where that language is used organically throughout the day.

Digital Tools and Resources for Communication in 8–12

To bring communication into the digital world for high school students, try creating podcasts.

Though podcasting never really went away, it has recently experienced a resurgence in popularity. A student-created podcast is a great way to share on a regular basis. At its very simplest, a podcast is just an audio recording posted online. Of course, you can also step up the learning opportunities by creating a more polished production. For that, you need some recording and editing tools as well as a way to organize and present your online podcast feed.

Recording Tools. Most computers today have basic microphones built in, and you can get acceptable audio quality from them, but it is probably worth investing in a better microphone if you're going to do recordings on a regular basis. A professional quality studio microphone will set you back several hundred dollars, but isn't necessary for classroom podcasting. You can get a decent mic for under $100.

Software. The best choice for basic podcast recording and editing on Windows is Audacity (http://audacity.sourceforge.net). It has many features and it's free. The interface takes some time to learn and use, but it isn't daunting. Audacity is also available on Mac, but if you are going to try to share your podcast through iTunes, then GarageBand (https://www.apple.com/mac/garageband) is still an excellent choice, although the most recent version removed some of the advanced podcasting features.

Sharing. To share your podcast, you can simply post the audio files to your class website and direct people to the links for downloading. I recommend, however, that you learn how to create an RSS (Rich Site Summary) feed of your podcast, which can then be submitted to Apple for inclusion in their iTunes podcast directory. The easiest way to do this is to create a blog to go along with your podcast. Wordpress is my tool of choice, and they have detailed instructions for creating a podcast blog (http://en.support.wordpress.com/audio/podcasting on Wordpress.com, or http://codex.wordpress.org/Podcasting for self-hosted blogs).

For More Information. The classic guide to podcasting in the classroom is Will Richardson's (2010) book *Blogs, Wikis, Podcasts, and Other Powerful Web Tools for Classrooms*, now in its third edition. Tony Vincent also has a great introduction to podcasting on the Reading Rockets website (http://www.readingrockets.org/article/creating-podcasts-your-students), and there are some good additional tips by Maya Payne Smart at Edutopia (http://www.edutopia.org/podcasting-student-broadcasts). Teachers considering podcasting should also review the Creative Commons Podcasting Legal Guide (http://creativecommons.org/podcasting).

TECHNOLOGY INTEGRATION FOR COMMUNICATION: CASE STUDY

Classroom Blogs

This case study comes from my own teaching experience as a teacher in the Centennial School District in Pennsylvania.

> I was the first teacher in my school to embrace using a blog with students. It was 2009, and my three principals were skeptical about posting student work on a public website. Why three principals? I was teaching gifted support classes in three schools within the district, so I reported to each on the days I was in his or her building. They all knew I had done my research, however, and after I described the benefits to the students, as well as my plan for anticipating and preparing for the possible risks, the principals gave me the collective go-ahead.
>
> The project was to plan a trip to Mars and involved students in Grades K–5 from all three schools working together asynchronously. There was a great deal of research and math involved in planning the trip, and students had complex problems to solve. Each problem began with a blog post that I created on the site.

The advantage was that all the students could respond with their suggestions, comments, and solution ideas, and everyone could immediately see what all the other students had posted.

I shared the blog address with parents, who also posted comments and were able to talk to their children about the problem at home. The students' classroom teachers and the school principals had access to the site, and although none of them posted comments, a few did tell me they read the blog and appreciated knowing what was going on in my class.

We even had two "virtual student teachers" from a Canadian university that semester. A connection of mine who taught at the university had his students working with teachers around the world via social media and other websites. The student teachers interacted with my classes, providing additional feedback, asking questions, and helping my students learn more about their chosen problems.

The blog became the focal point of all of our work that year, with links to reference materials and project resources, and instructions for the problems. The students were also much more deeply engaged in the project than my previous class the last time that I taught the unit without the blog. They interacted more with each other, and had tremendous opportunities to practice their written communication skills. Overall, the project was a great success, and the feedback from parents and principals was very positive.

FOR PLC AND STUDY GROUPS

1. Review the six levels of mastery from the beginning of this chapter. How many of them can you confidently say your students demonstrate? How can you design an assessment plan to evaluate not just whether your students can perform a skill, but how deeply they have mastered it?

2. Do all students need to demonstrate mastery at Level 6 for all concepts and skills? If yes, how do you make certain that they have the opportunity to demonstrate that mastery? If no, what is the minimum acceptable level of mastery for a classroom in the school or grade level that you teach? What if a student never reaches that level? What if a student already exceeds that level before you start teaching?

(Continued)

(Continued)

3. If you are an elementary teacher, chances are that you teach language arts in addition to math. How can you leverage the reading, writing, speaking, and listening skills you are teaching in language arts to help create a culture of problem solving all day? What are the natural crossovers that allow you to embed language arts instruction into your math time and vice versa? What aspects seem to be more of a forced fit?

4. If you are a secondary teacher reading this book, you probably specialize in teaching math. How can you broaden your own knowledge about language arts to improve your skill at teaching reading, writing, speaking, and listening in your math course? Can you collaborate with another teacher to develop activities, units, or even whole courses that blur the boundaries we create between disciplines in middle and high school?

5. Writing is time-consuming, and school schedules typically allot a limited time to math. What barriers exist in your situation to incorporating more writing about math? What creative solutions allow you to increase opportunities for math writing without changing your schedule? Alternatively, can you make an argument for increasing the amount of time you spend on math instruction each week? Where can the extra time come from and how can you persuade administration to make the change?

5 Collaboration

Learning to collaborate is part of equipping yourself for effectiveness, problem solving, innovation and life-long learning in an ever-changing networked economy.

—Don Tapscott, Business
Executive, Author, and Consultant

I collaborate and sometimes don't agree at all with my collaborators' opinions. It forces you to understand why you don't agree with something: what's the fight you're picking.

—Ira Glass, American Public
Radio Personality and Producer

BUT, ISN'T THAT CHEATING?

When I cotaught fifth grade math, one of our routines was to give our students a challenging problem every Friday. At the beginning of the year, I explained the expectations for the weekly problem.

"You have one week to work on it. Next Friday, you need to turn in your solution with all of your work."

The next bit of information often resulted in some confused looks. "During the week, we encourage you to check with each other, work together, and collaborate on your solutions. You can't submit a group response—everyone must turn in his or her own solution—but you can help each other work it out."

Once, early in the school year, one of the students named Puja responded, "You mean, you're going to let us cheat?"

"It isn't cheating if we're letting you do it," I responded. "Now, I don't expect you to take one person's answer and copy it. The material you turn in needs to be your own work. Everyone also has to understand it enough so that I could ask you to explain it all by yourself. But, feel free to work together when you're figuring it all out."

This was a culture shock for many of them, and the first time they did work together, it was as if they were testing us to see if we really meant it.

Puja approached me timidly. "Um. Hi. We did this together, and we both put basically the same thing down. Are you sure that's okay?"

"Did you write your own explanation in your own words?"

"Yes."

"If I asked you to solve the problem again and to explain your thinking, could you do it without your partner's help?"

There was a momentary silence. "Well," Puja said, "I think so."

"As long as you can do that, it's okay that you worked together."

We rarely have to solve problems alone, and we are rarely prevented from getting support from experts and colleagues. It is therefore important to encourage Collaboration as an important part of every mathematics classroom. To see why, let's look more closely at the first two stages of the problem-solving cycle.

RECOGNIZING, DEFINING, AND REPRESENTING PROBLEMS

In Chapter 2, I shared two examples of problems that highlighted the difference between typical classroom exercises and problems which required some more complex, DOK Level 3 application of ideas:

Exercise: Miguel collects baseball cards. Last week he had 217 cards in his collection. Today, his aunt gave him two dozen more for his birthday. How many cards does he have now?

Application: You and your friends are going to play a game using a set of cards numbered from 0 to 9. On your turn, you are going to draw three cards one at a time from the face-down deck. The object is to make the largest 2–digit number you can using your cards, and then the leftover card is discarded. The catch is that you must decide where to write each digit before you draw the next card: do you write it in the tens place, ones place, or discard it? If you draw a 4 as your first card, in what place should you write it, and why?

Both of these problems have their place in the math classroom, but a classroom based on the 5 Principles emphasizes Level 3 application as often as possible.

Both of these are examples of what we call *well-defined problems*. Well-defined problems have three main components, all of which are known at the start:

- **The Starting Conditions.** All of the information needed is available and clearly stated. Nothing is missing, and everything is clearly defined for us. In the first problem, we know how many cards are in the collection and how many are in the gift. For the second, we know what all the numbers are.
- **The Rules.** For most exercises, we know we can apply the standard mathematical operators. In the second problem above, the rules of the process are clearly defined: we draw three cards one at a time, writing the selected digit in one of the available spaces.
- **The Goal.** In a well-defined problem, we know the ending condition. In the first problem, we know we need to find the total number of cards. In the second, we need a plan for winning the game.

All well-defined problems can be reduced to an algorithm that mechanically produces a solution given the starting conditions, rules, and goal as inputs. This is one way to think about it: a computer can always be programmed to solve a well-defined problem. Some, like the baseball card problem, become trivial once you find an algorithm. Even the second problem can be solved with an algorithm, which becomes a winning strategy for the game. The chance element of this problem means the algorithm just maximizes your likelihood of winning rather than guaranteeing it, but there is still one best strategy to gain this advantage when playing, and a computer can be programmed

In most school math classes, the only problems that students encounter are well-defined problems. This is unfortunate, since most problems outside of school are ill-defined.

with this strategy. Some well-defined problems, such as playing chess, have extremely complex algorithms, but it is still possible to write computer programs that play chess as well as the best human players.

In most school math classes, the only problems that students encounter are well-defined problems. This is unfortunate, since most problems outside of school are ill-defined: that is, one or more of the three parameters are either incomplete or entirely missing. This is where the problem-solving cycle we outlined in Chapter 3 really begins to show its value. In particular, students need the opportunity to practice Steps 1 and 2 as listed below:

1. Recognize or Identify the Problem. The problem solver needs to determine that there is a problem to be solved. We often do this for students, because we use what Getzels (1982) calls *presented problems*. Although, to exercise this skill, we should sometimes use *discovered problems*. "Such a problem already exists, but it has not been clearly stated to the problem solver. In this case, the problem solver must put together the pieces of the puzzle that currently exist and seek out a gap in current understanding in order to 'discover' what the problem is" (Pretz, Naples, & Sternberg, 2003, pp. 5–6).

Unfortunately, this is very difficult to achieve in the classroom, since we have specific learning outcomes for students, and we don't have the luxury to just hope those discoveries occur naturally as we explore whatever problems arise in the course of the day.

But, we can simulate this experience of discovery for students. Consider the second problem above. Instead of presenting the entire problem as stated, let's restructure the lesson.

First, let's teach the game, but leave off the final question. Allow students to play this game for a while. If you have been cultivating the principle of Conjecture, then it's likely that someone will express frustration at their bad luck, or state that the game is unfair, or wonder if there's a strategy that will help them win. When this happens, stop the game, share the comment with the class, and discuss it for a few minutes. If it doesn't occur naturally, you can seed the conversation with a well-timed speculation of your own. "It looks like everyone is just playing their numbers randomly. I wonder if there's a way to make sure you have the best chance of winning no matter what cards you get."

Students now "discover" an interesting probability problem: is there a way to know what I should do with my first pick to improve my chances of winning? This query can turn into a general statement of probability ("If I pick a low number first, don't put it in the tens place") or you can extend it to a more in-depth statistical analysis of the possible outcomes. Either way, it requires the students to move on to Step 2 in the problem-solving process.

2. Define and Represent the Problem Mentally. With discovered problems like these, it is not enough just to ask the question. Now, students need to define exactly the goals and parameters of the problem.

In the case of the card game, the overall goal is clear: to win the game. But, what exactly does that mean? Students need to dig further and realize that winning involves some chance (we don't know what card will come up next) and strategy (you get to pick where each card goes). Students who recognize that the winning strategy is about probability of getting a high number for each pick and then using that probability to put the number in the most advantageous location are able to define the goal a little more specifically: to create a strategy that lets them choose the best spot for a given number without calculating the probability every time.

The parameters are both the rules of the game and the unstated consequences of those rules. Students must realize, for example, that once a number is picked, it can't come up again on their turn, but all numbers are available again for the next player. Students should also recognize that the winning strategy is not the one that guarantees winning in a specific round, but one that gives the best chance over time to collect points.

Finally, the students must represent the problem mentally. In this case, it means the student must categorize each digit based on the probability of selecting a larger or smaller digit, and thinking about it in terms of the location where it can contribute the most to the value of the final number without limiting the possibility of drawing another digit which would fit better in the slot selected.

Part of representing the problem also involves recognizing that one can simplify the problem in order to practice strategy building in a less complex situation. In the simpler version, the player chooses one card and decides whether to keep it or discard it. The player then selects a second card. If the first one was kept, the second one is discarded, and vice versa. The goal is to end up with the largest digit of all the players.

Few students will demonstrate these skills innately. They must be taught and cultivated, and they are easier to learn when students work and think together. The ideal environment for learning and practicing these skills, then, is one built on Collaboration, which is the third Principle of a modern mathematics classroom.

 Think Avengers, Not the Lone Ranger

MP1 and MP3

In a traditional classroom, students work alone, and the emphasis is on the skill fluency of an individual. Math in school has been treated as

> Big-picture innovation depends on our ability to share ideas and build on each others' work.

primarily a solo activity, but in the real world, problems are not solved by individuals acting alone. Collaborative learning processes and structures also contribute a great deal to the individual mathematics learning of students (Johnson, Johnson, & Stanne, 2000; Nunnery, Chappell, & Arnold, 2013; Stoner, 2004). The modern mathematics classroom is therefore all about the "we."

Collaboration also means de-emphasizing competition. Prevailing wisdom says that competition promotes better, more innovative thinking. This is, after all, the basis of capitalism and evolution. On certain scales, this works. But, in the long run, big picture innovation depends on our ability to share ideas and build on each others' work. Steven Johnson says that even Darwin depended on open access to ideas that came before him when he developed his idea of survival of the fittest. "Openness and connectivity may, in the end, be more valuable to innovation than purely competitive mechanisms" (Johnson, 2010, p. 21).

Sylwester (1995) talks about another important collaborative tool in the teacher's toolbox: storytelling. When we sleep, the brain spends about seventy-five percent of the time creating and maintaining neural pathways. The other twenty-five percent of the time is spent dreaming. Dreams are our brain's way of constructing meaningful and memorable stories out of the new pathways. Sylwester argues that we ought to consider a similar ratio in schools: spend at least a quarter of the time finding meaningful and memorable relationships between the ideas, facts, and skills we are teaching. "The best school vehicle for this search for relationships is storytelling as a broad concept that includes such elements as conversations, debates, role playing, simulations, songs, games, films and novels" (Sylwester, 1995, p. 103). Daniel Pink (2005), in his book *A Whole New Mind*, argues that storytelling is one of the most important skills we can hone in the twenty-first century. Stories help us "understand the world not as a set of logical propositions, but as a pattern of experiences" (p. 102). Pink also says that stories are how we connect with other people.

It isn't surprising that the collaborative element is important here.

> Because working together requires providing explanations, asking questions, listening, responding to others' perspectives, and giving feedback, learners are compelled to reorganize their thinking in light of the interchange with a partner. Partners also are given the opportunity to explore differences in their own versus others' knowledge in ways that would not happen in individual work. (Ostroff, 2012, p. 152)

So instead of encouraging your students to be Lone Rangers, think the Avengers. (Yes, I know I'm mixing my literary references!) Although each member of the Avengers can function independently and is an individual with his or her own strengths and weaknesses, as a team they are able to accomplish much more. And when you get right down to it, even the Lone Ranger worked with a partner.

Students should most often tackle problems in pairs or small groups. Solving problems solo should be a rare event in the classroom, and should only be practiced after students are confident in their problem-solving abilities in groups. Collaborative work also encourages and facilitates dialogue and debate, which are key elements of the Principle of Communication.

 ## Math: All Day, Every Day
MP2, MP4, and MP7

Just as we think of language arts as an "all day every day" thing, mathematical problem solving likewise needs to break out of its narrowly defined silo and filter into the rest of the school day.

Primary grades routinely teach these subjects together, and it's not until upper elementary that we tend to start building more walls between the content areas. Why not break them back down? Train yourself and your students to be on the lookout for mathematical connections throughout the day. If you teach other subjects, build explicit math references into your other lessons. Relate mathematical ideas to current news, and to events in the school and community. Have students bring in real world problems—even ones which aren't immediately math-like—and look for ways to apply problem-solving techniques.

Schools in Finland, often promoted as some of the world's best, are beginning to make this shift (Garner, 2015). Think about how breaking down the silos of subject areas can promote deeper collaboration and understanding for your students as you read the strategies in this chapter.

> Just as we think of language arts as an "all day every day" thing, mathematical problem solving likewise needs to break out of its narrowly defined silo and filter into the rest of the school day.

GRADES K–3

Collaborative activities in the primary grades need to be well-structured and carefully planned. Students need explicit instruction in the process of working together as well as in the problem-solving process and in math concepts.

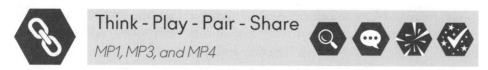

Think - Play - Pair - Share

MP1, MP3, and MP4

One strategy which gives structure to Collaboration is a reversal of a traditional teaching approach known as "I do, We do, You do." In that technique, teachers first demonstrate a skill. Then students practice it with guidance, and finally practice it independently. Instead, we should *begin* with individual exploration, followed by group work, ending with whole class discussion. Magdalene Lampert, longtime math teacher and professor of education, calls this "You, Y'all, We" (see Green, 2014).

I've tweaked this strategy a bit for primary grades, borrowing some terminology from Lyman's (1981) collaborative learning structure called Think-Pair-Share. In my version, there are four stages to the structure:

1. **Think**. Students are presented with a problem and spend a minute or two thinking about it. It can be helpful to prompt thinking with specific questions to help focus students:

- What is the question we have to answer?
- What do we know?
- Is there anything we need to know?
- What kind of answer should we get?
- What could I do that might help me get an answer?

2. **Play**. In this added step, individual students are encouraged to just experiment and play with the problem. There is no expectation that any student will solve it on his or her own. This is free play, intended to let them try things out and see what happens. Give it plenty of time, especially when students are just learning the structure. They may become frustrated. Avoid too much prompting or specific guidance at this point. Instead, reassure them that there is nothing they can do wrong during this step. Everything can help them understand how the problem works. Some questions you might find useful here are:

- What did you notice when you tried that?
- What would happen if you tried the opposite?
- What do you think a math teacher would try next?
- What is the silliest thing you can think of to try? Did you try it?

3. **Pair**. Now students work together. Use pairs or small groups depending on the task. They should continue their explorations, sharing whatever they learned while playing individually. Now, though, they should shift gears and seek a complete solution to the problem. Set an expectation that

groups may not consider a problem "solved" until everyone in the group is satisfied with the result and is comfortable explaining it to someone else.

4. **Share**. Finally, engage in a whole class discussion of the solutions. Compare solutions between groups. Look for similarities and differences, and talk about why one or more of the solutions works and why others may not. During this stage, you can introduce algorithms and techniques to make solving similar problems more efficient in the future.

Why bother with this approach? Isn't it faster and easier to directly instruct and just teach the skill? A 2004 study by Klahr and Nigam seems to say it is, and much of our traditional approach is based on that assumption. Teach it, and they will learn. Klahr and Nigam compared a group of students who were taught a specific skill through direct instruction to another group who learned it through inquiry. They found that there were students in both groups who mastered the skill, but the comprehension of the direct instruction group was significantly higher, even when students were asked to apply the skill to a new context.

However, Dean and Kuhn (2007) examined this again a few years later, looking not just at the effect of each type of instruction, but also how well mastery was retained over time. They looked at three groups: one learned the skill only through an inquiry approach over ten weeks, one received direct, explicit instruction in the strategy, and the third had the direct instruction lesson followed by the inquiry-based practice. They then administered two assessments: one which was a direct test of the skill, and another which tested transfer to new contexts. These two tests were repeated five weeks later to assess retention.

In the Dean and Kuhn study, students in the direct instruction-only group had the poorest performance on every assessment. The direct instruction plus practice group did the best on the tests immediately after instruction, but five weeks later, the inquiry-only group performed the best.

The authors are quick to note that this is not an *efficient* way of teaching. Nor is it enough to continue to work with one problem or type of problem. "[P]ractice, however, does not achieve its maximum effect after 12 sessions working with a single problem. The new content introduced at the second assessment shows some suggestion of reinvigorating the engagement of the [inquiry] group" (Dean & Kuhn, 2007, p. 395).

If our goal is for students to gain skill quickly and efficiently, then direct instruction is a clear winner. But, if we want students to learn skills that they can retain, and use, and grow throughout their lives, we must consider the less efficient approach.

> If we want students to learn skills that they can retain, and use, and grow throughout their lives, we must consider the less efficient approach.

Digital Tools and Resources for Collaboration in K–3

Even young children should have the opportunity to extend their networks outside of their immediate classroom and school. Digital technology allows us so many opportunities to connect students to their counterparts in other parts of the world. One powerful tool for this is Skype. The *Skype in the Classroom* project (http://education.skype.com) illustrates several ways teachers can use Skype in the primary math classroom:

- **Search the math category** for teachers with similar needs or interests: https://education.skype.com/categories/math. Possibilities include challenging each other to solve problems your students created, or to expand the "Think - Play - Pair - Share" strategy and find a problem that both classes can work on together.

- **Play the Mystery Skype game** (https://education.skype.com /mysteryskype). This involves connecting with another classroom in another part of the world, and having the two groups of students ask questions of each other until they are able to determine the exact location of the other classroom. While this tends to focus on geography, why not give it a math spin by answering questions with a math problem that the other class has to solve to decipher the clue?

- **Invite mathematics professionals** to speak to your class and help them learn about how the math they're learning is used in the real world. This isn't limited to the guest speakers on the Skype website. Scour your own professional network for people who can add excitement and interest to your math lessons.

- **Go on virtual field trips** to interesting places around the world. Visit https://education.skype.com/partners for a list of some of the organizations that are working with Skype to create virtual experiences for students. Many of the projects have opportunities to explore math concepts, which certainly can provide examples of problem solving in the real world.

GRADES 2–5

Math Workshop
MP1, MP3, MP4, and MP5

Teacher Erin Klein uses a math workshop approach in her classroom. "The idea of giving ownership of learning back to the student is an important aspect of our learning environment," she says. "Students appreciate having autonomy and being able to explore subjects in ways that fascinate them." One feature of a math workshop that Klein occasionally uses is to have students invent their own problems.

In younger grades, this can be as simple as providing students with an answer (11 apples) and asking them to work in groups to come up with two different, interesting problems that could have that answer as the solution. As students learn how to do this, at first by trial and error, and later by reasoning backward, it gives them insight into the problem-solving process.

In the older grades, give some constraints or provocations. For example, ask students to respond to one or more of these challenges:

- Create a problem that has at least two different yet correct answers.
- Create a problem that uses the information in a given infographic.
- An archaeologist a hundred years from now finds a scrap of paper from a student's notebook with the last part of a solution on it. See if you can recreate the problem it came from and complete the solution. (Provide students with the last step or two from their own work on previous activities, or from student work in prior years or other sections.)
- Make a "problem-finding box" containing various objects with mathematical potential: measuring cups, old calendar pages, seashells, keys, menus, magazines, partly full containers of paperclips or peppercorns, food labels, sheet music, pages from a car owner's manual, etc. Have students select two random objects to give to another team and challenge them to create an interesting and meaningful problem that uses both objects.

Klein believes that because a math workshop turns over the responsibility for finding and figuring out their own problems to students, it is a powerful way to spark their curiosity and keep them engaged. Though it is harder, the effort required by both teacher and students is worth it. "It is our job as educators to cultivate that passion and find ways to shape the curriculum to the individual," Klein says. "Math workshop allows for more personalized learning and creative problem solving with collaboration amongst peers."

At least once a week, model your math classroom around the same methods you use during a writing workshop. Give students interesting problems to solve, and let them work on their own projects at their own pace, helping and supporting each other in small groups, doing mini lessons with students who need

"Math workshop allows for more personalized learning and creative problem solving with collaboration amongst peers." (Teacher Erin Klein)

additional support or guidance. Build problem-solving centers where they can work in shifts. Stock the classroom with lots of mathematical puzzles and games and let them play.

Digital Tools and Resources for Collaboration in 2–5

There are a number of online services which allow you and your students to collaboratively share and edit documents. **Google Drive** (http://drive .google.com) is probably the best known and the most ubiquitous, and it is an excellent choice for any classroom, especially if your school uses Google Apps for Education (https://www.google.com/work/apps/education).

Another tool is **Padlet** (http://padlet.com). Formerly known as Wallwisher, this tool allows you to create a shared wall where anyone can post information. It has some features which make it particularly useful for supporting all-day math in the intermediate grades:

- **No login required**. Anyone with the link to your Padlet can edit it by default, though it is hidden from the public in searches. Students do not need to have an account to use it.
- **Easy link names**. When you create a Google Doc, the link to access it is extremely complex and impossible to remember. With Padlet, however, you can create meaningful links that are easy to read and remember. For example, the Padlet associated with this book is at http://padlet.com/geraldaungst/5CMath.
- **Multiple options for privacy**. You can control the restrictions on your wall—how public or private. Do you want to create something the whole world can see? Make it totally public. Do you want more control over who is viewing or editing? You can protect it with a password or add moderation so that new posts are approved by you before they appear on the wall.

You and your students can use Padlet in a number of ways to encourage more collaboration in the classroom:

1. Create a general Q&A board for students to post questions that you or others can answer. Share the board with parents and professionals to get more ideas and help.

2. Create a "to-do" board for larger projects and group work. Padlet even has a 3–column backdrop to organize notes into "To-Do," "In Progress," and "Done" categories.

3. Record a timeline of ideas and notes by using the "Stream" layout to organize all notes in the order they were posted.

4. Make a Padlet center where you and students can post ideas for problems to solve and then they can post their solutions as they finish them.

5. Post a separate Padlet for each problem and have students post their ideas for solving as well as their completed solutions. Students can then read and discuss or critique each others' solutions.

GRADES 4–7

By fourth grade and beyond, students are capable of solving more complex, longer-term problems in math, which require more time, more thought, and more difficult reasoning. These kinds of problems are perfect for collaborative solving.

 Cultivating Hunches
MP3, MP5, and MP8

Consider this prompt:

Lockers in our school are assigned by homeroom, and homerooms are assigned alphabetically. Students with last names starting with "A," therefore, get the low numbered lockers on one end of the hallway, while students with last names starting with "Z" get the high numbers at the other end of the hallway. As a result, some students have to walk a lot further to get to their lockers between classes.

Is there a better way to assign lockers so that everyone has a short route to their locker and can easily visit it several times a day to avoid carrying a heavy backpack?

There is a great deal of math involved in the solution to this problem, including geometry of the space and passing time between classes. It also allows for a very wide variety of possible solutions, depending on the sophistication of the student solving it.

But, it also requires some subtle reasoning. Students are likely to look at this and immediately have an idea about best solution options. However, the best ideas to complex problems are rarely generated from these gut reactions.

The snap judgments of intuition—as powerful as they can be—are rarities in the history of world-changing ideas. Most hunches that turn into important innovations unfold over much longer time frames.

They start with a vague, hard-to-describe sense that there's an interesting solution to a problem that hasn't yet been proposed, and they linger in the shadows of the mind, sometimes for decades, assembling new connections and gaining strength. (Johnson, 2010, p. 77)

Certainly we don't have decades in which to allow problems to germinate in the classroom. Nevertheless, we can still encourage the idea of letting ideas simmer for a while without moving immediately to a solution. Johnson suggests the idea of a "public hunch database" where humanity could compile all of the fragments of ideas in one place for others to browse and contemplate.

> Imagine a huge whiteboard in the classroom where students can jot down their thoughts, no matter how small, about a problem everyone was working on.

What if we adapted this idea on a smaller scale for the classroom? Imagine a huge whiteboard in the classroom where students can jot down their thoughts, no matter how small, about a problem everyone was working on. Or for each problem, a large sheet of butcher paper that could be spread out on a table or hung on the wall during problem-solving sessions? You could also create a binder where each page or section was devoted to a single problem and students could continually add their ideas and their comments about others' ideas.

Digital Tools and Resources for Collaboration in 4–7

Although analog methods are certainly viable, digital technology is ideally suited to compiling, accessing, and organizing students' hunches.

A simple **Google Doc** or **Padlet** can suffice, or perhaps an online form with a spreadsheet to log ideas and comments so they can be grouped and sorted easily.

Tricider (http://www.tricider.com) can make the whole process simpler, however. You can post a problem on the site along with one or two ideas to prompt student thinking.

Students can then add their own ideas, including vague hunches or even questions of their own. For each idea, others can add an argument or discussion and mark it as a pro or con. They can also vote for the ideas they think are best, and the ones with the most votes rise to the top of the list.

Here's an example I posted on the Tricider site using the locker problem from earlier: http://www.tricider.com/brainstorming/2m0nmF1Xbet. In order to get a feel for the tool and how it works, feel free to add your ideas, vote for existing ones, and add arguments. Another interesting tool for collating and organizing hunches is **Trello** (https://trello.com). Trello's main structure is the *board*, which then consists of any number of *lists*. Users can add cards into each list, drag them from one list to another, and comment on them. For example, lists can categorize ideas about solving a problem in various stages, from vague hunches, to viable plans, to verified solutions. As the ideas develop, students can move them to the appropriate category list. You can also create a list for disproven ideas, which can provide valuable insights from analyzing all of the ideas that didn't work.

Because cards can also contain links, images, and attachments, Trello can be a one stop shop for organizing the supporting information involved in solving a complex problem. In addition, cards can have checklists and due dates, so students or teachers can manage the project right in the Trello board.

GRADES 6–9

 ## Daily Data
MP2, MP4, MP5, MP7, and MP8

For several years, I taught a unit on data analysis that I called "Surveys & Polls." Not a terribly catchy title, but the students enjoyed it because I gave them the opportunity to investigate questions they were curious about. Each group came up with a topic they wanted to poll, developed their poll questions, conducted a survey, and analyzed the data they collected. Commonly they would poll their friends about their favorite cafeteria foods, popular clothing brands, and so on.

One year, a group decided to do a survey about the Motion Picture Association of America (MPAA) movie rating system. They asked students and parents about whether the ratings were useful and helped them to make decisions about which movies to watch. Although I no longer have the results of the survey, I recall that they generated some interesting data, specifically about the differences in perceptions between kids and adults. In particular, kids and parents differed on the value of the PG-13 rating. Parents thought it was helpful, while kids thought that the restriction for age thirteen was too high and that the majority of younger kids could see those movies without a problem.

After some discussion, the students decided that they wanted to do something more meaningful with the data than just make a poster and share it with the class. They wrote a letter to Jack Valenti, who at the time was the president of the Motion Picture Association of America, sharing their data and making suggestions about how to improve the system. Several weeks later, the group received a personal letter from Mr. Valenti thanking them for sharing their results and telling them the MPAA would consider their suggestions.

This was a powerful lesson for the class about how understanding and organizing data can help people make decisions and support arguments. It was also a tremendous opportunity to practice the Principle of Collaboration. The group came up with much more interesting ideas by working together than if they had worked alone. They also were able to see the data in different ways and talk about different interpretations. The group process of figuring out how to present and share the analysis also deepened each individual student's understanding of the math concepts involved throughout the process.

> Students should have daily opportunities to work with data, analyze it, organize it, and present it.

Students should have daily opportunities to work with data, analyze it, organize it, and present it. Language arts teachers often include a "daily edit" activity as part of their classroom routine. To provide a daily opportunity for collaborative work in math, try presenting a "daily data" activity:

- Grab an interesting statistic from the news, a book or magazine, or one of the data sources listed in Figure 5.1. Share the statistic with the class and have them work in groups to provide an interpretation. Take a few minutes to share, discuss, and argue the interpretations.

- Share a statistic as above, but present an existing interpretation, either from another source, or one you generated. Ask students to work together to determine if the interpretation is valid or if it suffers from a typical misuse of statistics. (See Figure 5.2 for a list of misuses of statistics.)

Figure 5.1   Sources of Data for Classroom Use

- U.S. Government Open Data website: http://www.data.gov
- U.S. Bureau of Labor Statistics, K–12 Education Resources: http://www.bls.gov/k12
- Data on children and families in the United States from the Annie E. Casey Foundation: http://datacenter.kidscount.org
- Drexel University Math Forum list of data sets: http://mathforum.org/library/topics/data_sets
- Hope College Department of Mathematics list of data sets: http://www.math.hope.edu/swanson/statlabs/data.html
- Middle School Data Resources from National Council of Teachers of Mathematics (NCTM): http://www.nctm.org/profdev/content.aspx?id=11688
- American Statistical Association, Useful Websites for Teachers: http://www.amstat.org/education/usefulsitesforteachers.cfm
- Weather data from Weather Underground: http://www.wunderground.com/history or the National Oceanic and Atmospheric Association (NOAA): http://www.ncdc.noaa.gov

Figure 5.2   Misuses of Statistics

- **Bias and Manipulation**. Samples may be deliberately small, or chosen from a non-representative group. For example, polling only wealthy people about the economy would not produce a useful statistic.
- **Unnecessary Qualifiers**. Korn (2014) points out that sports announcers use this frequently when announcing that a player has just set the "*team* record for most yards gained *from scrimmage by a running back in the first quarter*." By choosing such a specific statistic, with the qualifiers in italics, it makes the achievement seem more impressive than it actually is.
- **Using the Wrong Statistic**. Mean, median, and mode are all useful for describing a set of data, but they have different meanings in different contexts. Harvard Business Review (2000) gives the example of a poll where the mean rating was 4, but every individual rating was either a 3 or a 5. Since no one in fact gave a 4 rating, the mode, or in this case modes, are a more meaningful average to use. Zaccaro and Zaccaro (2010) give the example of a murder trial where the defense said, accurately, that only 1 in 2,500 abused women are killed by their abuser. What they did not share is that when abused women are murdered, the abuser is the killer ninety percent of the time.
- **Assumptions**. A poll reports that ninety-nine percent of all audience members enjoyed a particular musical. When you dig further, you discover that this is based on the fact that one percent of the audience requested a refund. The assumption that everyone who didn't ask for a refund therefore enjoyed the musical is not necessarily the case.
- **Amplification**. Exemplified by claiming a "thirty-three percent increase in sales" when in fact only four items were sold this month compared with three last month.
- **Cherry-Picking and Arbitrary Comparisons**. Researchers who are trying to prove a point might run an experiment ten times and only report the one that has favorable results. They may also collect data and then set or ignore arbitrary parameters after the fact that make their results come out the way they want. For example, according to the National Aeronautics and Space Administration from 2004 to 2011, the five-year running average Land-Ocean Temperature Index did not change (NASA, 2015). This seems to contradict climate scientists' warnings about our effect on the environment. But, the Index doesn't tell us the temperature; it tells us how *fast the temperature changes*. If you look further back, you can see the more dramatic acceleration. I merely selected two particular years to represent the start and end of the range, which made the index appear not to change.
- **Misleading Graphs**. Truncated graphs with an axis that doesn't begin at zero, or pictographs where the images are different sizes, create false visual impressions that lead to incorrect interpretations of data.
- **Correlation Is Not Causation**. Causation is very difficult to prove experimentally. Correlation is much more commonly reported. Be careful of assumed correlation, also. Just because two things both took place at the same time does not imply that they were related.

Source: Types of misuses compiled and adapted from Gelman & Nolan, 2002; Korn, 2014; Wallis & Roberts, 1962; and Zaccaro & Zaccaro, 2010

- Gather longitudinal data by taking a minute or two each day to record a few interesting facts about the school or your students. For example, have someone time the morning announcements, or keep track of the different breakfast foods your students ate that day. Once you have a few months worth of data, start looking for patterns and trends. Are Mondays different than Fridays? What happens on the day right before or after an extended break?

- Practice simple data literacy skills. Huff (1982) suggests several questions we can use to "talk back" to statistics and make sure we understand them accurately. All of these are excellent group discussion questions for students to ponder together.

  ○ Who says so? We need to know the source of the information and potential biases.
  ○ How do they know? The method of collection and organization is critical.
  ○ What's missing? The presentation must be complete and rigorous.
  ○ Did somebody change the subject? Data can be misused to support a claim about something subtly different.
  ○ Does it make sense? Think through the implications of a statistic before you accept it as given.

Digital Tools and Resources for Collaboration in 6–9

If you are getting into the habit of compiling small amounts of data every day and assembling it into data sets for later analysis and discussion, it's important to have a place to capture it all and organize it for that later work.

While individual shared documents are useful, and in particular Google Sheets provide a collaborative place to collect useful data, there are other tools which expand your students' ability to collect and discuss their project data in one location, including **Evernote** (https://evernote.com), **OneNote** (http://www.onenote.com), **Simplenote** (http://simplenote.com), and **Google Keep** (https://keep.google.com). We'll focus on just Evernote, but the others have some similar features depending on your needs.

Evernote is free to use. Evernote's files are organized as "notebooks," which are essentially folders to collect different kinds of notes and files. Notebooks can be shared with as many other users as you like, making them an excellent place for students to compile their ideas and data for later use and analysis. Because Evernote allows multiple types of information and files to be gathered into one notebook, students can begin to see "data" as more than just numbers. Now, data can be any source of information, particularly when they can search it all within one space.

Teacher Nicholas Provenzano of Grosse Pointe Public Schools, Michigan, uses Evernote extensively with his students. He finds the shared notebooks to be a key feature in facilitating student collaboration. "What that does is it gives kids the opportunity to not have to work at exactly the same time as each other. My kids will work on projects together, and when we didn't have access to Google at all, it allowed them to work at their own pace."

Evernote is also device agnostic (meaning it does not require specific hardware to run and will work on almost any device). Provenzano, who was the International Society for Technology in Education's 2013 Teacher of the Year, frequently finds his students pulling out their phones to snap a quick photograph of some notes on the interactive whiteboard, sending them to Evernote, and adding their class notes there. They also use their devices to collect research and data from whatever sources are convenient. Evernote has browser plugins that allow students to capture data from a website and annotate it. Students can store everything in a shared project notebook, tagging it to facilitate organization. Later, Evernote's powerful search features allow them to locate anything they've saved anywhere in the program.

While Google Apps allow students to collaboratively create in real time, Provenzano likes that Evernote lets them collaboratively *curate*, which is a different skill. "My students use the web clipper and Skitch to collect and annotate websites and screenshots. They take pictures right from their phones and upload them directly to Evernote. They can easily move and share information."

GRADES 8-12

In high school, students should be regularly working together to solve rich, complex problem scenarios that require them to build mathematical models and apply the skills and concepts learned during instruction.

 Model-Eliciting Activities (MEAs)
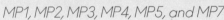
MP1, MP2, MP3, MP4, MP5, and MP7

Dr. Richard Lesh, retired professor of Learning Sciences and Mathematics Education at Indiana University, has developed activities explicitly designed to prompt students to construct and use mathematical models, apply mathematical thinking, and reveal their thought processes.

Here's one example of what Lesh calls *model-eliciting activities*, or MEAs. Students are first given a newspaper article about a contest where competitors construct paper airplanes to win one of two awards—"Most Accurate" and "Best Floater." The problem-solving teams then get the following problem statement along with a data table:

> In past competitions, the judges have had problems deciding how to select a winner for each award (Most Accurate and Best Floater). They don't know what to consider from each path to determine who wins each award. Some sample data from a practice competition and a description of how measurements were made have been included.

To make decisions about things like being the best floater, the judges want to be as objective as possible. This is because there usually are only small differences among the best paper airplanes—and it seems unfair if different judges use different information or different formulas to calculate scores. So, this year, when the planes are flown, the judges want to use the same rules to calculate each score.

Write a brief 1– or 2–page letter to the judges of the paper airplane contest. Give them a rule or a formula which will allow them to use the kind of measurements that are given in Table 1 to decide which airplane is: (a) the most accurate flier, and (b) the best floater. Table 1 shows a sample of data that were collected for four planes last year. Three different pilots threw each of the four planes. This is because paper airplanes often fly differently when different pilots throw them. So, the judges want to "factor out" the effects due to pilots. They want the awards to be given to the best airplanes regardless who flies them. (Lesh & English, 2001)

Note that the problem here is not to determine a winner for *this* set of data, but to develop a rule or model that will work for *any* set of data in any airplane contest with similar parameters.

It's also important to recognize that the goal here isn't for students to come up with the "right" model, or even the best model, but just to generate *one* model that they can justify and which leads to further discussion and refinement based on that discussion and feedback.

Digital Tools and Resources for Collaboration in 8–12

MEAs are a rich and extensive strategy for the high school math classroom, and detailed instructions for using or creating them is well beyond the scope of this book. Thankfully, Purdue University has created a website with many of these MEA problems along with thorough resources for teaching with them. You can access these Case Studies for Kids at https://engineering.purdue.edu/ENE/Research/SGMM/CASESTUDIESKIDSWEB/index.htm.

TECHNOLOGY INTEGRATION FOR COLLABORATION: CASE STUDY

Trello: Collaborative Organizing

Rik Rowe teaches mathematics at Wilmington High School in Massachusetts. He uses a number of digital tools to help his classes collaborate and communicate with each other. This year, he began using Trello (http://www.trello.com) to allow his classes to collaboratively track their learning progress.

Since our students spend much of their time on their mobile devices, I'm thinking it's wise to enable them to access their learning on those same digital tools. I've been in search of a platform that is browser-based, collaborative, intuitive, clutter-free and accessible from laptops, tablets, and smartphones. Trello enables us to create a "Flow of Learning" on a clutter-free canvas where we can collaborate and communicate with our students regardless of where any one of us is located. We have the ability to integrate Trello with Evernote, Twitter, IFTTT (a web-based service that allows users to create "recipes" that allow different digital tools to share information and automate tasks) and more tools that I rely on daily. I've recently created a Flow (an IFTTT recipe) from Evernote or Email directly to any Card on any List on any Board. This is helping to have Trello be a central repository of content for our learners.

Our students easily adapted to Trello and are already communicating with me and each other on the Trello Cards specific to our content. I'm attaching documents and links so our students can operate from a central location to learn, understand and apply our content. Trello enables us to create a left-to-right flow that becomes intuitive on our learning journey. As we make progress through our learning, we move our Cards through the Lists from "Still to Learn," "Now Learning," "Learning To Be Assessed," "Assessed Learning" and "Reflections of Learning." The "Reflections of Learning" List holds a card entitled Reflections document. I see this as a shared document for all learners to record our thoughts and reflections on our learning. I can see it housing many "aha" moments and after the fact realizations. All learners can access these thoughts at any time, but might be especially useful to study for tests.

We are in a Standards-Based learning environment and our Trello Cards are created based on the Standards we're learning. Cards can easily be added, moved and archived to create our Flow. Our students and I are already accessing Trello on our mobile devices to work through suggested practice, review checklists aligned to our material, see relationships with Mind Maps and have conversations with each other to clarify and deepen our learning. Each student is a Member of the Board and can move their avatar to any Card indicating that content is something they're currently learning. All individuals in our learning environment can see at any time who is where, who might be helpful as a resource and who might need our support.

With each passing day, I'm realizing there is so much more to Trello to track our learning journey and to keep all involved persons connected to each other and our learning.

FOR PLC AND STUDY GROUPS

1. Two of the most common objections to doing more group work, particularly in math, are that the better math students tend to carry the load, and there is no individual accountability for mastery of skills and concepts. What are some ways to emphasize individual accountability without sacrificing the culture of Collaboration in your classroom? When is it appropriate to ignore accountability? Are there times to de-emphasize collaboration? What do you believe is the appropriate balance?

2. Several of the strategies in this chapter involve implementing daily routines that help you establish a collaborative environment. Review the daily routines you have now and plan ways to revise them to shift toward a 5 Principles approach. What daily routines are most conducive to this revision? Which ones are more complex to adjust? Are there school routines and practices that either promote or restrict collaboration in the classroom? What can you do about these?

3. Though teachers ordinarily only have one school year with a particular group of students, consider the possibility of developing large, complex problems and data sets that classes could work on for years. How can you leverage technology to carry problems over from one year to the next? What could students do to build on the knowledge and hunches of previous classes? What possibilities does this open up for interesting learning experiences?

4. What can you do to make more room for discovered (rather than presented) problems in your instruction? How could you ensure that even with discovered problems you are meeting curriculum expectations and standards? What complications arise if you use more discovered problems, and how can you anticipate ways to resolve them?

5. Since the 5 Principles are intended to be a framework, begin thinking about how you might integrate Conjecture, Communication, and Collaboration into one problem or activity. What opportunities lend themselves to such integration?

6. Consider identifying a school or community problem for students to tackle. How can the 5 Principles apply to such a process? What math opportunities can arise from this kind of problem? What are the risks of this approach, and how can you mitigate them?

6 Chaos

Much of formal education . . . feels like learning the pieces of a picture puzzle that never gets put together, or learning about the puzzle without being able to touch the pieces.

—David Perkins, Founding Member of Project Zero,
Harvard Graduate School of Education

One must still have chaos in oneself to be able to give birth to a dancing star.

—Friedrich Nietzsche,
German Philosopher, Critic, and Poet

FEATURES OF EFFECTIVE MATH INSTRUCTION

In a survey of research about how classroom teaching methods can affect student learning, Hiebert and Grouws (2007) find only two features of teaching that have enough evidence to confidently say they have a significant impact on student conceptual understanding of mathematics. The first is when teachers and students explicitly attend to concepts (as opposed to merely discrete facts and skills). The second is when students struggle with important mathematics. Both of these are embedded in this cultural principle of Chaos.

Imagine that Edison found a filament for the electric light that worked on the first try. What did he learn about filaments? Not much, except for knowing one way that works. If at some point in the future, he tried the filament in a bulb with a different configuration and it didn't light, he'd be at a loss. Each time Edison "failed," however, he added a new piece to the puzzle of what makes a good filament, and what makes a poor one. Each failure contributed new information to his growing database, allowing him to refine his process until he had a solution that worked well.

The same thing happens for students. Remember that the goal of solving a math problem is not for students to get good at solving *that* problem. It is for them to get better at solving *all* problems, including ones that no one has thought of yet.

Steven Johnson (2010) talks about the kinds of environments that promote better problem solving and the innovation we talked about in Chapter 1.

> Innovative environments are better at helping their inhabitants explore the adjacent possible, because they expose a wide and diverse sample of spare parts—mechanical or conceptual—and they encourage novel ways of recombining those parts. Environments that block or limit those new combinations—by punishing experimentation, by obscuring certain branches of possibility, by making the current state so satisfying that no one bothers to explore the edges—will, on average, generate and circulate fewer innovations than environments that encourage exploration. (p. 41)

Johnson goes on to concisely express the rationale for this fourth Principle with this statement: "Good ideas are more likely to emerge in environments that contain a certain amount of noise and error" (p. 142). For this innovative climate to happen in our modern mathematics classroom, it's gonna have to get messy.

DISORGANIZED CHAOS

In my early days and months as a teacher, chaos ruled my classroom. I had not yet learned how to successfully manage a group of preteens to enable learning to take place. Or at least the learning I wanted to happen.

I spent much of my first three years studying classroom management strategies and techniques. My conversations with colleagues and principals often revolved around how to get kids to "stay on task" and how to maintain an orderly classroom. I established and enforced rules, expectations, and consequences for misbehavior.

It definitely helped cut down on the chaos, and it helped preserve my sanity. In retrospect, though, I'm certain that my motivation was more about ensuring compliance than enhancing learning. If you had the misfortune to be in that fourth-grade class at Burnett Elementary School during the 1991 to 1992 school year, I sincerely and profoundly apologize.

Math class needs to be a place with a healthy dose of chaos. I'm not talking about the kind of chaos my students experienced that first year. I'm talking about the messiness of real problem solving. Thomas Edison famously said, about the path to finding the right filament for his light bulb, "I have not failed. I've just found 10,000 ways that won't work."

> We can't expect kids to learn to cook without making a mess in the kitchen.

One of the biggest problems with math instruction is that we show students the one way that someone else figured out would work, and then ask them to rehearse it to perfection. We never let kids experience the ten thousand ways that won't work. This sounds counterintuitive. After all, if we want kids to be good problem solvers, don't they need to be able to get to correct solutions to the problems? Definitely. I'm not talking about aimless, endless anarchy. But, we can't expect kids to learn to cook without making a mess in the kitchen.

 ## It's Gonna Be Messy
MP1, MP2, and MP5

There is a misconception about math that because it demands precision, it is also neat and clean. While we want a clean final draft, math—like writing—is more about the process than the product. But, also just like writing, the process is the opposite of clean.

We tend to want students to proceed in strict linear fashion from the start of a problem or exercise through to its solution. However, real world problems don't work that way, and math class problems shouldn't either. "Problem solving does not usually begin with a clear statement of the problem; rather, most problems must be identified in the environment; then they must be defined and represented mentally" (Pretz, Naples, & Sternberg, 2003, p. 3). There will be false starts, reversals, and abandoned trails when students start to really solve problems.

This is hardly a new idea in math instruction. Japanese researchers have been studying the effects of an open-ended strategy since the early 1970s (Shimada, 1997). And while I often hear teachers talking about using open-ended problems, I rarely see the kind of messy problem solving that leads to sophisticated understanding of mathematical ideas.

PRODUCTIVE STRUGGLE

Understand that we are not talking about a free-for-all classroom. Perhaps a more precise description of the messiness we are looking for is *the edge of chaos*, a phrase coined by Christopher Langton, which Steven Johnson (2010) explains is "the fertile zone between too much order and too much anarchy" (p. 52).

Langton uses a metaphor of the phases of matter to help explain where this zone lies. The most chaotic state is a gas: molecules are in constant motion, and there is little hope of any patterns forming due to its extreme volatility. The opposite state is a solid: stable, predictable, and rigid. There are lots of patterns, but they are predetermined and unchanging. It's unfortunate that the word "rigor" tends to evoke this state, since little learning takes place here.

> The sweet spot in learning, this liquid state, is what Jackson and Lambert call "productive struggle."

However, in a liquid state, the molecules have the benefits of both extremes: there is lots of motion, but the motion can be contained and directed. New patterns can form, but they can also be retained.

The sweet spot in learning, this liquid state, is what Jackson and Lambert (2010) call "productive struggle." To understand this, let's look first at its opposite—Destructive struggle. This condition

- leads to frustration;
- makes learning goals feel hazy and out of reach;
- feels fruitless;
- leaves students feeling abandoned and on their own; and
- creates a sense of inadequacy.

It is certainly out of a desire to avoid these responses that we often feed students the algorithms and clues to getting to a solution. Don't be in too much of a hurry to bail them out, though. Let them get lost in the muck for a while. It's okay. By reasoning through and working it out on their own, they begin to better understand the territory where the problem lives. Mason, Burton, and Stacey (2010) tell us, "Probably the single most important lesson to be learned is that being stuck is an honourable state and an essential part of improving thinking" (p. viii). This is where productive struggle comes in. Productive struggle is a condition that

- leads to understanding;
- makes learning goals feel attainable and effort seem worthwhile;
- yields results;
- leads students to feelings of empowerment and efficacy; and
- creates a sense of hope.

Th _____ ouws (2007) envisioned.
Withir _____ tand their own thought
proces _____ ter solutions.

Br _____ ," but I prefer to call it
Chaos _____ collective attitude and
approa _____ hers and students who
can lea _____ ictive struggle produce
more s _____ 7) report on a study by
Inagak _____ hen classroom discussions ' _____ conjectures along with
correct _____ the discussions and
improv _____ ough both their verbal
statem

[handwritten note: Productive vs. Destructive Struggle]

Teaching Like Video Games
MP4, MP5, MP7, and MP8

Video games give us a nice model for building the kind of skill and confidence that allows students to engage in productive struggle. In video games, it is rare to have an instruction manual; players simply dive in. The early levels provide easy experiences designed to orient the player to the game world, introducing concepts one or two at a time. Yet, these levels are never merely exercises. Accomplishments make a difference in the game, and progress is always relevant to the overall game goals. In spite of the fact that they are typically simple and the stakes are low, the player never feels like he or she has to wade through practice material to get to the "real" game.

Another feature of these games is that the introduction of new skills is almost always done by putting the player into a real world (or in this case, game world) situation where the skill is necessary. Often, players are given minimal procedural directions and have to work out, in the context of the game, how to apply the new skill and ways to use it effectively. In early levels, there is often an obvious choice, but it is still left to the player to make it. The game quickly takes the player to novel situations where the obvious application of a technique may not work and the player has to reason his way to a solution.

Games respect the Chaos needed for the learning process by lowering the stakes involved in making mistakes. A character's death, for example, is rarely permanent, and even when it is, it is usually possible to go back to a previous checkpoint to try to regain a position. The ability to attempt the same puzzle over and over, learning from each attempt, just as Edison did, is one of the things that makes games fun and engaging.

Math class can be this way too. David Ginsburg promotes this in his work coaching teachers.

> How can we teach students to make sense of problems and to persevere when a problem is challenging? In my experience, we can't. We can, however, *create classrooms that cultivate these skills*, and the formula for doing this is simple: provide students challenging problems, and empower them with the skills and resources they need to solve those problems. Then get out of their way. (Ginsburg, 2014, para. 3, emphasis mine)

So what would a math class look like that is based on this video-game model?

First, provide students with a series of problems of increasing complexity. Begin with simple, straightforward problems where the solution is relatively obvious and the student can succeed easily. Quickly move on to problems that are more challenging, until students are working on very rigorous, deep, and open-ended problems around the same general topic. Keep the content challenging as well. Don't assume young children can't handle the material; advanced concepts and content should be a part of math instruction from the first day children walk into Kindergarten (Claessens, Engel, & Curran, 2014; Engel, Claessens, & Finch, 2013).

Two additional elements characterize the video-game approach from a more traditional one: feedback and pacing.

Two additional elements characterize the video-game approach from a more traditional one: feedback and pacing. After each attempt, successful or not, provide students with specific feedback about their performance. Briefly, point out areas they should examine again, places where they made errors, and opportunities for improvement. Guide them toward a more successful performance, but avoid pointing them directly to a solution.

The second element, pacing, is driven not by the curriculum guide, but by the student. In a video game, as soon as the player successfully demonstrates a skill, he or she can move on to the next level. In the classroom, as soon as the students successfully solve a problem, allow them to move on to the next level of complexity. While this takes additional planning and record keeping, and students need some help navigating the new process where they set their own pace, it is well worth the investment when you see students who are able to attack difficult problems with enthusiasm and self-confidence.

 ## Avoiding Filters
MP3

It is tempting to come to a struggling student's rescue immediately. After all, it's in our nature to want students to succeed, and when we see struggle, we want to help. Remind yourself, though, that too much help is no help in the long run. It creates dependence and prevents growth.

> Too much help is no help in the long run. It creates dependence and prevents growth.

Instead of pointing students directly to the correct solution path when their first attempts fail, initially focus on what worked, then on *why* their solution did not ultimately succeed. Suggestions are always better than corrections, and use thought-provoking, reflective questions instead of leading questions. For example, in a multi-step problem where the student selected a good strategy, but made a computation error, instead of pointing out the mistake, or asking whether they checked their work, try the following series of questions:

What kinds of errors could have caused this? Which one of them is most likely for you? Which one is the simplest to find and fix? Try looking there first. Of course, it is entirely possible that they select and look for the wrong kind of error. Let them. When they struggle, come back and ask further questions: Did that path work? What other kinds of errors did you think of? Did you try looking for those? Are you keeping track of the areas you're checking so that you don't have to go back and look again?

You may be thinking that it is better to give the student a checklist of steps to follow when solving and checking errors in their problems. Again, resist this urge. Let the student discover that they are going through the same steps over and over, and that it would be helpful to have a checklist. Then, either help them to create their own (which is the most powerful option), or offer to provide one that you found helpful to your own problem solving. Either way, now each student will internalize the value and application of the checklist in a way not possible if you had provided it first.

I suspect you may also be writing some objections in the margins. *This is all well and good, but it's not how things work in the real world.* Leaving aside the question of how much and how soon the process of learning needs to precisely mirror the world beyond school, let's consider how product development works at one of the world's largest and most successful companies, Apple Inc.

In a traditional manufacturing process, each stage is handled by a separate team. The design team begins by developing the concept for a product. The engineering team then determines how to make it work. The manufacturing team works out the processes for building it, and then the marketing team figures out how to sell it and how people can buy it. "This model is so ubiquitous because it performs well in situations where efficiency is key," says Steven Johnson, "but it tends to have disastrous effects on creativity, because the original idea gets chipped away at each step in the chain." For example, engineering decides they can't make the entire product work as designed, so they filter out the parts they can't do. Manufacturing takes out some other elements that are too complex or too expensive to mass-produce.

Apple, on the other hand, puts all of these teams in one room and works on all stages at once.

> The process is noisy and involves far more open-ended and contentious meetings than traditional production cycles—and far more dialogue between people versed in different disciplines, with all the translation difficulties that creates. But, the results speak for themselves. (Johnson, 2010, p. 171)

The bottom line is that Apple's approach, just like the Chaos principle, avoids limiting thinking and promotes creative problem solving. Students learn to honor each other's ideas. They also learn to have confidence in their own brainstorms, and eventually to be able to assess and revise them, without intervention by experts. Once they get to handle truly unsolved problems, where there is no predefined solution and no expert to guide them, they are able to jump in and persist until they arrive at a useful solution. Don't sacrifice quality learning for efficient teaching.

GRADES K–3

The Principle of Chaos may seem simple to achieve in the primary grades, due to the high energy level of most K–3 students. Teachers can take advantage of the natural Chaos in their students, just as they use their natural curiosity. The key is making the chaos productive instead of random. A great place to start is by emphasizing the importance of the process.

 ## The Journey Is More Important Than the Destination
MP3, MP4, and MP8

Teacher José Vilson (2012) uses five strategies in his classroom to build a culture of purposeful Chaos:

- Allow more mistakes
- Support their struggle
- Let the kids teach, too
- Answer a question with more questions
- Personalize the questions

"We should strike a balance between using direct instruction and exploration," Vilson says, "leaning more on the exploration piece" (para. 5). In other words, we need to allow more students time to experience the journey instead of always focusing on the destination.

When I was about nine, I went to Cub Scout camp for the first time. Every day, a group of us got on a bus and we rode for an hour or so to Camp Delmont. Saturday was parent day, and my dad was taking me. So we hopped in the car, and Dad said, "Tell me which way to go."

Now, I had ridden in the middle of the bus and knew vaguely (at best) which way we had gone. I did recall one of the other kids commenting at one point that we were getting on the Turnpike. Or was it the Expressway? No, Turnpike, definitely. "Go to the Turnpike." We barely left our neighborhood when Dad took a left at an intersection where I was absolutely certain that the bus had driven straight. "No, Dad, go straight!" So, he calmly got turned around and back onto the route I remembered.

It wasn't long before I was completely lost. But, I wasn't about to let Dad know, after my absolute certainty about the first turn. So, he kept driving, and I kept directing him as best I could. "Are you sure you drove through Norristown?" he asked. "Yep, Dad, I'm sure. Right through here. Yep."

Miraculously, we managed to end up at the camp in time for Parent Day. In retrospect, Dad,

> When we hand students turn-by-turn directions, and tell them to just follow them until they get to the end to miraculously obtain the right answer, we are really handicapping them.

as the king of maps, probably already knew where the camp was and how to get us where we needed to go.

The point is that when we hand students turn-by-turn directions, and tell them to just follow them until they get to the end to miraculously obtain the right answer, we are really handicapping them. Inevitably, the problem comes when the students have to make the journey on their own. If they can recall all of the steps in the right order, they're fine. But, if they miss a step, or make a wrong turn, they'll be just as lost as I was trying to get my dad to Camp Delmont.

Instead, we need to give students the tools to plan their own journey. Give them a map and teach them how to use it. Help them understand the specific techniques and strategies (the roads and highways) for solving math problems.

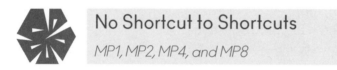 

No Shortcut to Shortcuts
MP1, MP2, MP4, and MP8

We have a tendency in the primary grades to focus on product for a few reasons. First, the concepts we are teaching are, to us, very simple and straightforward and don't warrant extended time. Remember that, for young students, these are sophisticated ideas, and that the exploration time is critical to understanding.

One way to help students focus on the process rather than the outcome is to provide them with both problem and answer. Ask them to plot a route that successfully moves them from start to finish. Then, to emphasize process even more, instead of collecting and grading the work, which merely replaces one destination with another, ask them to share their solutions and explain to their peers what they meant and why their solutions are good.

Another strategy that helps build problem-solving skills, and embraces the principles of Chaos, is to leave problems unresolved overnight. Deliberately give students a problem which you know cannot be solved in a short period of time. When you get to the end of your math class, students expect you to reveal the answer. Don't. Let them sleep on it.

John Medina (2014) shares a study that looked at this very idea. Researchers gave students math problems to solve, which had a hidden shortcut the students could discover. Twelve hours later, they were given a similar set of problems. Students who slept during those twelve hours were three times more likely to discover the shortcut than those who didn't. Sleep allows our brains to continue processing information, and it allows us to continue working on problems.

Digital Tools and Resources for Chaos in K–3

To help your young students focus on process instead of final product, try taking photographs and video of them working on problems. Use **Animoto** (http://animoto.com) to build a video slideshow documenting students in action, and have them share the videos with parents and peers, describing what they were doing at the time.

Students can also collect photos of different people solving the same problem. Have them show the different paths they took to get to the destination with narration to explain how the paths were the same or different.

GRADES 2–5

In the intermediate grades, students are more capable of self-monitoring and self-regulation, which allows you more freedom to set up an environment that promotes Chaos. Building on students' understanding of process, we can introduce non-routine problems that are more complex and engaging than the routine problems we find more frequently in published materials.

We often see people promoting the use of real-world problems in math, and I used to be one of them. But, mathematician and educator Dr. Gordon Hamilton tells us that this is the wrong adjective to use.

> Real-world mathematics discounts the possibility of using fantastic imagery that's already in an elementary student's mind: Star Wars, Tolkein, Greek mythology. None of these are real-world, and yet all provide a rich backdrop to your mathematical problems. But, we can go further. We don't need to have stories behind some mathematics problems. I've had fantastic success with some abstract mathematics.

> So how do I choose what's a good problem and what's a bad problem? It's got nothing to do with real-world, or abstract, or Star Wars. It has to do with "engaging." A quality mathematics problem has to engage a wide spectrum of student ability. (Hamilton, 2011, 5:47)

In short, a good problem for mathematics instruction is one that has a low entry ramp, accessible to learners with a limited or weak skill set, and a high ceiling that can engage the strongest math students. Many real-world problems can do this, and using them is an important part of good math instruction. But, we shouldn't limit ourselves and avoid some potentially powerful problems just because they aren't "real-world" enough.

Non-Routine and Unsolved Problems

MP1, MP2, MP3, MP4, MP5, and MP7

So, we accept that things might get messy in a problem-solving classroom. Let's explore some specific strategies that create a "safely" messy environment and that take advantage of the Chaos to build problem-solving skills.

The ability to solve novel problems is an important element of the shifts in Common Core mathematics. In the past, we tried to prepare students for assessments by showing them examples of every possible problem they might see. A better strategy is to give students frequent opportunities to attempt non-routine problems that stretch their problem-solving muscles and give them strategies for what to do when they aren't certain what to do next.

Sawada (1997) describes some of the advantages of using these non-routine problems:

Advantages

1. Students participate more actively in the lesson and express their ideas more frequently.
2. Students have more opportunities to make comprehensive use of their mathematical knowledge and skills.
3. Even low-achieving students can respond to the problem in some significant ways of their own.
4. Students are intrinsically motivated to give proofs.
5. Students have rich experiences in the pleasure of discovery and receive the approval of fellow students.

Let's look at a specific example to see how it supports mathematical reasoning and productive struggle.

Scott the Painter

Consider the following problem, from Rudd Crawford's wonderful collection he calls "Stella's Stunners," a website resource from the Ohio Resource Center.

Scott the painter is going to paint the floors of 16 rooms. He is in room #1. He can go through any door (see diagram [in Figure 6.1]). He cannot return to any room once he has painted it—slow-drying paint! In what order should he paint the rooms? Obviously, #16 must be last.

Figure 6.1 Scott the Painter

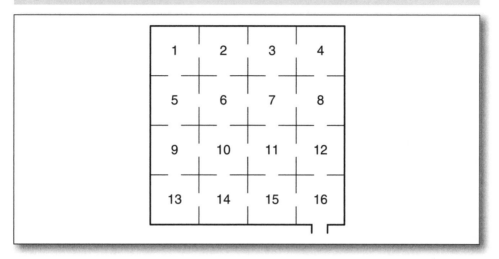

(For more problems like this one, including a link to Crawford's entire collection, see the Digital Tools and Resources for Chaos in the Grades 2–5 section later in this chapter.)

Before you read on, take a few minutes and try to figure it out yourself. It's important that you actually do this and don't read straight through to the solution, because the experience will help you grasp the importance of this concept. You may even want to share the problem with some colleagues or students and see what you can come up with.

Ready? Did you find a solution?

Typically, people who first encounter this problem will attempt multiple pathways from room 1 to room 16. No matter what route you take, however, there will always be at least one room unpainted (or one room with footprints in the fresh paint). The instinct is to conclude it is not possible to paint all 16 rooms and exit from room 16.

But, don't leap to this conclusion too quickly. There is a solution which will allow all 16 floors to be painted and still keep Scott's shoes clean. Read the problem again. Scott begins in room #1. Put yourself in his place. Imagine you are actually standing in this unusual building and need to paint the floors. How would you proceed?

Have you solved it? Not yet? Perhaps this last hint will unravel it for you: Scott the Painter begins the problem in room #1. Must he paint that room first?

As first stated, without any clues, and with the assumptions that you probably brought to your initial attempts, this is very much a non-routine problem. There is no procedure or algorithm you learned at any time during your school math class that you can apply to solve this.

Heuristics

What we can use, though, are what mathematicians call *heuristics*. These are thinking strategies rather than computation techniques that can tease apart the threads of a difficult problem the very same way Maniac Magee unraveled Cobble's Knot in Jerry Spinelli's wonderful 1990 children's book. Some of the common problem-solving strategies we see in school textbooks (and which are usually given light treatment at best) are heuristics: draw a picture, work backward, guess, and check. In this case, the heuristic we need to use is to identify our assumptions—meaning any conditions we started to work with, but which aren't actually stated in the problem—and then change them. We assume that since the problem says Scott begins in room #1, he must start painting there.

> Heuristics are thinking strategies that can tease apart the threads of a difficult problem the very same way Maniac Magee unraveled Cobble's Knot.

Once you realize there is nothing to prevent Scott from walking around before he lays down any paint, he is suddenly free to begin painting in any of the rooms, which opens up a great number of paths to a solution.

Problem Solving Is the Best Test Prep

An objection I often hear to the use of more non-routine problems is that it takes time away from preparing students for the ubiquitous high-stakes tests that pervade schools today. We feel pressure to abandon a problem-solving culture, even if only for a few weeks of the year, and replace it with test prep.

There is a serious and dangerous assumption behind this belief, however. Our approach to preparing students to perform on most tests, including those we devise ourselves, is to make it so that every problem they might encounter on the test is simply an exercise, or at worst, a routine problem.

To understand why this is a problem, let's put it in a musical context. Think of the state test as a sight-reading challenge. Sight-reading is when a musician is presented with a brand new piece of music never before seen and is asked to play it with no rehearsal. The musician is allowed a brief amount of time to review it silently and think about it, but the first time he or she plays it, it counts. This is routine in music evaluations, and is a standard part of the audition process.

Our normal approach to test prep, if applied to music, would be like providing students with a thick book full of every musical phrase they might see in a new piece and practicing it until it is perfected. The hope is that when the students see the sight-reading piece, they will be able to recognize every phrase and reproduce each one flawlessly. If the pool of possible pieces is small, this strategy might work. But, the analogy quickly escalates out of control if we want the students to be prepared for any possibility.

So, is it hopeless? Of course not. Instead of practicing the individual phrases, we teach students about patterns and theory in music, and ask them to practice selected specific scales, rhythms, and intervals that will be useful. We also teach them how to identify difficult passages, and how to analyze and understand what they see on the page in order to accurately produce the music when they play it.

This is exactly the power of non-routine problems. By teaching students heuristics, instead of algorithms, we give them tools that enable them not just to solve specific kinds of problems, but to embrace and tackle any problem that might come their way, even if it is one they have never seen before. Instead of frustrating a student, teach about typical entry points that could be fruitful, and how to respond when an approach doesn't work out.

Non-routine problems give us a powerful way to introduce, teach, and practice various heuristics without worrying about driving immediately toward the correct answer.

Kindergartens Everywhere

Aviva Dunsiger is a teacher at the Dr. Edgar Davey Elementary School in Hamilton, Ontario, and a 2013 recipient of the Prime Minister's Award for Teaching Excellence. Her previous school experienced a problem which she saw as the perfect opportunity to get her students involved. The real-world experience is an excellent example of how Chaos can lead to better problem solvers:

> Our school was one of the last schools in the Board to get Full-Day Kindergarten, and we had a big problem: where were we going to put the additional classroom? Which location would lead to the least disruption to other classes and the least amount of construction? Where could we get a classroom that was big enough for 30+ Kindergarten students? Why was this location the best?

> With the help of my principal, Paul Clemens, and vice principal, Kristi Keery-Bishop, my [fifth grade] students were invited to help solve this problem and determine a solution that could be presented to the Board. They needed to consider the area and perimeter of the different classrooms. Unfortunately, many rooms were not perfect

squares or rectangles, which made calculating the area far more difficult. Some students took to a grid, while other students tried to split the classrooms into rectangles, and add up the different parts.

They quickly realized areas that they forgot to measure. Then they remembered about the need for a bathroom, and they had to start subtracting from their total area. There was constant remeasuring, recalculating, working through problems, and then determining a solution that tended to lead to more questions and more problems. In the end, none of the solutions were one hundred percent perfect, but the "chaos" of a real-world problem with much potential for math learning, helped students truly understand area and perimeter and the impact of both.

Digital Tools and Resources for Chaos in 2–5

For more non-routine and open-ended problems, as well as more ideas to promote deeper thinking and the principle of Chaos, visit any of the resources and articles below.

- **All Bets Are Off Outside Math Class:** http://www.edutopia.org/blog/all-bets-off-outside-math-class-matt-levinson
- **Art of Problem Solving:** http://artofproblemsolving.com
- **Bedtime Math:** http://bedtimemath.org
- **Dan Meyer's Three Act Problems:** http://threeacts.mrmeyer.com
- **The Educational Value of Math Puzzles:** http://www.edutopia.org/blog/recreational-educational-value-math-puzzles-deepak-kulkarni
- **MathPickle Unsolved Problems:** http://mathpickle.com/unsolved-k-12
- **Problems of the Month** (organized by CCSS standards categories): http://insidemathematics.org/index.php/tools-for-teachers/problems-of-the-month
- **Problems of the Month from Inside Mathematics:** http://www.insidemathematics.org/problems-of-the-month
- **POW: Problems of the Week** (links to various problem collections): http://www.math.com/teachers/POW.html
- **Problems of the Week from the Math Forum @ Drexel University:** http://mathforum.org/pows
- **Recreational Math:** http://www.math.com/teachers/recreational.html
- **Stella's Stunners,** over 600 non-routine math problems for all levels: http://ohiorc.org/for/math/stella

GRADES 4–7

Intermediate grade students are ready to pursue more of their own investigations in addition to solving problems presented to them. While our earlier chapter on Conjecture encouraged students to generate interesting questions and problems to solve, this proposed culture comes to full fruition when layered with Chaos, since each student can now pursue individual passions and interests, and we can let them branch off in directions that initially may not seem like math. For that, we introduce "Genius Hour."

Changing Scenery With Genius Hour

MP1

Most of my writing is done either at my desk or at my local Starbucks. Something about each environment brings out the best in my ideas and productivity. However, as I write this particular section, I'm sitting on a lounge chair in my back yard, shaded by the neighbor's tree, and surrounded by green grass (well, green*ish*; we could use some rain), chirping birds, and the smell of mint from our herb garden.

I make a point to change my location and scenery from time to time since it sparks new ideas and insights, and breaks me out of my rut. It's the same reason I used to take my students out of the classroom. When students are stuck on a problem, give them ways to change their environment. It could be as simple as having work spaces in your classroom set up in different configurations, with comfortable furniture and cozy lighting. Or allow them to go to other classrooms, the library, or even a hallway. Take the whole classroom on a "field trip" to work in the auditorium, cafeteria, or soccer stadium.

Besides physical changes in the environment, it can be helpful to try academic ones. Many teachers are beginning to embrace the idea of "Genius Hour," adapted from Google's "20% time." Google saw the value in allowing their employees to use up to a fifth of their work week on projects of their own choosing. Why not bring that innovation into your classroom?

Teacher and technology specialist A. J. Juliani has been a pioneer with Genius Hour, using it in his own teaching since 2012. His book, *Inquiry and Innovation in the Classroom* (Juliani, 2014), is a tremendous resource for developing the principle of Chaos through this practice. He says

> The Standards for Mathematical Practice are basically telling you to do genius hour. Here's the way to think about it: we have skills we want our students to learn and master, and we have assessments

where we want students to demonstrate that learning and mastery. In between there, we have content through which we want students to learn the skills. In the past, we had one source of content—a textbook for example—that we used to teach the skills. Now, content is interchangeable in many different areas. Students can read articles, watch videos, ask people, try things. Genius hour is giving students choice for resources, and also what kind of assessment they want to do. (A. J. Juliani, personal communication, July 17, 2014)

Consider taking one day a week in your math classroom to allow students to pursue any problem or mathematical question they want to explore. Give them some unstructured time to play, experiment, and research. Keep a running log of your observations: What practices do you see students using? What mathematical concepts and skills are they applying? Take photographs or video of students solving problems, and collect samples of their work. At the end of the process, give them opportunities to share and explain what they did, even if the work is incomplete or their solution doesn't quite work.

Digital Tools and Resources for Chaos in 4–7

One way to spark some interest in Genius Hour and give students ideas for what to explore is to connect to their interests. Earlier, we discussed how we can model math instruction after the learning progressions in video games. We can also bring the games that students play directly into the mathematics classroom. Video games lend themselves well to mathematical problem-solving applications, so try allowing students to experiment with their favorite games during Genius Hour. Two possibilities to consider are Minecraft and World of Warcraft.

Minecraft is a popular video game among teenagers. Challenge students, for example, to design a fireworks show in Minecraft then compute the most efficient way to craft all the Firework Rockets (http://minecraft.gamepedia.com/Firework_Rocket) needed for the show. The end result of their work should be a complete recipe, including a raw materials list and steps to construct the show.

World of Warcraft (WoW) is a massively multiplayer online role-playing game (MMORPG). If you aren't familiar with this type of game, it is essentially an online version of Dungeons & Dragons with millions of players from around the world playing and interacting together. Teachers Lucas Gillespie and Craig Lawson (2010) have developed a curriculum unit around this game called WoW in School: A Hero's Journey. At the project website, http://wowinschool.pbworks.com, the authors and their collaborators Peggy Sheehy and Helga Brown have numerous resources for integrating WoW into other content areas, including math.

Other video games also lend themselves well to mathematical problem-solving applications. Ask your students what they play at home, and have them look for connections to the math work you are doing in class.

GRADES 6–9

In the middle grades, it's important to help students to know when and where messiness is appropriate, and to learn how to tolerate it in their own work. There are times when precision and accuracy are needed, and even when they are working on a messy problem, there eventually comes a point where they need to clean it up. But, we don't want them moving there too quickly, or to avoid the messy work simply out of a desire to take shortcuts and get things finished.

Creating self-contained and focused problems that still require the Chaos of messy problem solving is a fantastic way to serve all of these needs.

iWonders
MP1, MP2, MP3, MP4, and MP8

To help free students from their dependence on right answers, give them problems that you don't know the answer to, or even problems that cannot be answered precisely. This begins to show middle-schoolers that not every problem has been solved already, and that learning math is not just about finding the answers that someone else already knows.

I call these problems **iWonders**, since a good way to generate them is to start with the phrase, "I wonder. . . ." These problems sometimes require a bit of data collection or research as well, which teaches additional important problem-solving skills. Here are a few examples:

- iWonder . . . what is the weight of all the asphalt and concrete on I-95?
- iWonder . . . if we took all the eggs laid by all the chickens in Iowa on May 4, 2010, how big an omelet would it make?
- iWonder . . . if Citizen's Bank Park were completely filled with popped popcorn, how long would it take the Phillies to eat it all?

> To help free students from their dependence on right answers, give them problems that you don't know the answer to, or even problems that cannot be answered precisely.

The point here, of course, is not to get to the "right" answer, but to have students justify their solutions and critique each other's work. Once they try a few of these, students quickly begin generating their own.

If we extend this into fantasy territory, it opens up deeper possibilities for argument and evidence since there's little chance students can find "real" data to feed the problem. Here are a few ideas:

- iWonder . . . how many pounds of food Bilbo and all the dwarves ate on their trip to the Lonely Mountain?
- iWonder . . . how many lightsabers could be powered by the energy generated at Hoover Dam . . . or perhaps how many Hoover Dams it would take to power one lightsaber?

Digital Tools and Resources for Chaos in 6–9

Looking for unusual sources of information to spark intriguing iWonder discussions in your classroom? Try some of these (with the usual reminder that they are not necessarily intended for young people and should be used with discretion):

- **Useless Facts:** http://uselessfacts.net
- **Listverse:** http://listverse.com
- **Worldometers:** http://www.worldometers.info
- **Tree of Life Project:** http://tolweb.org/tree
- **Flowing Data:** http://flowingdata.com
- **The Superhero Database:** http://www.superherodb.com
- **So Who Wins: The Any Character in the World Battles:** http://www.sowhowins.com
- **Mental Floss:** http://mentalfloss.com
- **Spurious Correlations:** http://www.tylervigen.com
- **100 Interesting Data Sets for Statistics:** http://rs.io/100–interesting-data-sets-for-statistics

GRADES 8–12

By high school, students should be pursuing intriguing problems of their own choosing on a regular basis. They need frequent opportunities to spend extended time on rich, complex, Level 4 problems, both individually and in groups. For high school students, there are many ways to extend to a deeper level the strategies already shared.

Dan Meyer, an advocate for better math instruction also shares many examples of how to renovate traditional textbook problems to make them more intriguing and less mechanical. You can see the archive of these "Makeover Monday" problems at his blog: http://blog.mrmeyer .comcategory/makeovermonday.

Here, I describe a different kind of strategy for promoting Chaos in high school math classes.

 ## Computer Programming
MP2, MP4, MP5, MP6, MP7, and MP8

Computer programming is perhaps the ultimate expression in the digital technology world of this principle, and it's a perfect place for high school students to apply their sophisticated problem-solving skills. An argument can be made that coding skills should be part of the core curriculum, and some schools are doing just that. If yours is not, however, why not consider building it into your math instruction?

Complete beginners can start with languages like **Scratch** (http:// scratch.mit.edu) or **Alice** (http://www.alice.org), which provide easy entry into the world of programming while still giving students the power to execute complex activities.

For a simple project with potential for very deep math, challenge students to translate the game of Nim into a computer program. Nim is an extremely simple game, and learning to play should only take a few minutes. Here's how the basic version works:

Nim is a 2–player game.

1. Begin with 15 counters or objects in a pile between the two players (see Figure 6.2).

2. Either player can go first.

3. On your turn, you may remove one or two counters from the pile, as demonstrated in Figure 6.3. It is not necessary to keep track of these; once out of play, the counters are irrelevant.

4. You may not skip your turn.

5. The player who removes the last counter from the pile wins. It does not matter how many counters you removed during the game.

Figure 6.2 The Starting Position for NIM

Figure 6.3 Two Students Playing NIM

Once students are comfortable with the game, which only takes a few minutes, challenge them to work together with their partner to identify any patterns in the play, and to develop a winning strategy. There is a guaranteed winning strategy, which I leave to your students to discover.

However, a deeper analysis reveals patterns to the game, which can be generalized into mathematical models. Even very young children can do this—I've successfully taught Nim to first graders and had them complete the pattern analysis. It is possible to delve much deeper into an analysis of the game and its variations (some of which you can find at http://nrich.maths.org/1209), but another productive path is to begin turning the game into a computer program. Scratch is ideally suited for this, since it's fairly easy to build a graphic interface, as shown in Figure 6.4, without learning a lot of complex programming. Here are the stages of development I recommend to allow a wide range of abilities to dive into the project at their appropriate level:

1. Build an interface that allows two human players to play the basic version, essentially just simulating the play environment and rules of the game.

2. Add a component that allows the computer to play against a human, following the rules and detecting when the game is over.

Figure 6.4   A Version of Nim Programmed in Scratch

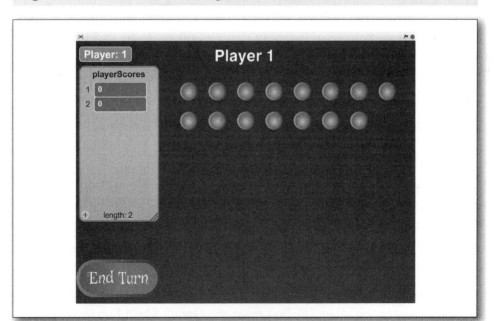

Source: https://scratch.mit.edu/projects/16135815/

3. Program the computer player to be able to win the basic version, using the strategy deduced during the analysis.

4. Program the computer to simulate other variations based on player choices (i.e., different starting position, different move options, and different winning condition).

5. Program the computer to win the game regardless of the variation chosen by the human player.

6. Program additional modes of NIM (options here are nearly limitless; online searches for NIM provide many varieties that students can explore, or they can invent their own versions).

Digital Tools and Resources for Chaos in 8–12

Since the main strategy in this chapter was itself a digital application, I conclude here with just one brief additional resource.

For a unique twist on programming, which may appeal to students who are more language-oriented, check out Interactive Fiction (also known as "IF", pronounced "eye-eff"). The text-based games and stories are filled with engaging puzzles, and the best IF requires a lot of messiness as students attempt to figure out what they can do with different items and features of the environment. Middle- and high-school students can even learn to author their own IF stories using the programming language **Inform** (http://www.inform7.com); the website contains a number of useful resources for learning to play and write IF, and for teaching with Inform.

TECHNOLOGY INTEGRATION FOR CHAOS: CASE STUDIES

Math Forum Digital Resources

Jacquelyn Confer-Sullivan is an elementary math specialist at the Cheltenham Elementary School in Pennsylvania. She describes how the Math Forum digital resources help her to support Chaos in her work with students who struggle with math and need additional support.

Throughout the school year, I informally incorporate all of the Mathematical Practices into our lessons as much as possible. The Problems of the Week (PoWs) from the Drexel Math Forum help a great deal, and allow me especially to let students experience the messiness of problem solving.

This week, my 4th grade students happily dove in to PoW #2827 and demonstrated the characteristics of successful learners that we have been striving for all year! After a brief assessment to ensure they had correctly interpreted the situation (they had), they enthusiastically engaged in the task. During this time, the kids discussed the nuances of the problem, reflected upon their understanding as it quickly evolved to higher levels, and employed various strategies to find the solution. I must say that it is EXTREMELY difficult for the control freak in me to NOT make suggestions! But I resisted, and I was quite impressed to see one child make the leap from Novice to Expert through his own reflections. He went back and forth several times with different strategies and expressed many "Ahhhhhhs!" when discovering more efficient methods to solve. These authentic, successful experiences have dramatically increased his confidence and refreshed his interest in learning.

This particular problem is excellent for Mathematical Practice 1— **Make sense of problems and persevere in solving them**. One student, who is typically reticent, needed more time to fully process the details of the problem and develop a strategy for solving. Though initially difficult for her, rather than give up and avoid the work, she persevered and navigated a course toward an answer. While her original course was murky and circuitous, after considerable reflection of her laborious and/or unsuccessful attempts, she adjusted to a much more efficient strategy. Despite not finishing (there was no "time limit" for finding an answer—our pull-out support time was over and they had to go back to their classrooms), she was thoughtfully engaged during the entire session. She left feeling proud of her accomplishments. The significant personal growth she is demonstrating as a learner is evident. She approaches the Chaos of problem solving with openness and flexible thinking rather than her previous "deer in the headlights"-like apprehension. I am so fortunate to be able to see these incredible changes!

I encourage taking safe risks, emphasize that problem solving is normally messy, and to embrace our mistakes as learning opportunities. The PoWs are EXACTLY what I needed to demonstrate and support my philosophy!

Inform 7: Interactive Fiction

Brian Francis Smith is a gifted support and English teacher at Cheltenham High School in Pennsylvania. He frequently makes interdisciplinary connections with other content areas and brings problem solving into his creative writing course.

As a high school Creative Writing teacher, a good deal of my job consists of being, well . . . creative. For that reason, I'm always on the lookout for opportunities to open my classroom door to visitors, innovators, and anyone with a story to tell. So when a colleague of mine suggested that he teach my Creative Writing class a mini unit on Interactive Fiction, I said, "Come on in."

On the day of the lesson, I arranged for my students to meet in the computer lab. As they filed in, a bumbling collection of good-natured eleventh and twelfth graders, they did not encounter an assignment or task. Rather, we greeted the students, directed them to an Interactive Fiction game website, and encouraged them to simply—play.

Over the course of the 47–minute period, I witnessed students laughing, pondering, collaborating, and breaking off from the crowd. I heard exhales of frustration and claps of high-fives. I watched as a group of 16- to 18–year-old students, from a multitude of backgrounds and ethnicities, remembered something they had forgotten: how to learn while having fun. When the focus of the lesson shifted from playing games to creating games, I saw introverts become extroverts and followers become leaders.

Was the lesson plan clean and neat? No. If you happened to pass by my Creative Writing class on that particular morning you would have observed a group of teenagers laughing, rolling around on their chairs, and bouncing from computer to computer. In other words, you would have witnessed a mess. But, what was really happening under this perceived veneer of chaos? My students were actually learning about the precision and accuracy of language required when writing fiction for a computer game. They were learning about the value of struggling through to a solution when trying to solve a complicated problem. They were learning about collaborative writing, a technique infrequently emphasized in an education system that commonly sees writing as an individual sport. They were learning about fun. And innovative thinking. And creativity. And how oftentimes the three are seamlessly intertwined.

As the bell rang and one of my students shouldered her book bag out into the hallways of high school, I wondered about her next class. Would it be the old classroom of scripted rules and staid routines? Or would her next class allow for freedom, creativity, and yes, even a little messiness? I hoped it would be the latter. Because once you've experienced a little chaos in the classroom, it's hard to go back.

FOR PLC AND STUDY GROUPS

1. In this chapter we explored ways to incorporate video games into mathematics instruction. Gamification of learning is a popular topic in some education circles. What other benefits can game-based design bring to math instruction in your school or classroom? What are the downsides? There is likely to be misunderstanding, if not opposition, from some stakeholders about game-related teaching. Anticipating the problems, how can you prepare students, parents, and fellow educators to learn about and demonstrate the educational value of these techniques?

2. A culture of problem solving and the Standards for Mathematical Practice are central to the premise of this book: that the main mission of K–12 mathematics should be to develop students into confident and secure problem solvers. But, teachers are still expected to "cover the content" detailed in the rest of the standards. How can you teach all of the required content and skills without compromising the Principles of a problem-solving culture?

3. Another recurring theme of this book is the importance of emphasizing effectiveness over efficiency. How can you insulate your students and yourself from the intense pressure to cram "efficiently" for the state test and stay the course of effective development of a problem-solving culture? Conversely, if you must comply with specific mandates related to test preparation, particularly in the weeks leading up to the exam, can you meet the expectations while still honoring the Principle of Chaos as described in this chapter? Are there creative options for accomplishing the requirements and also nurturing innovative thinking?

(Continued)

(Continued)

4. Administrators who walk into a classroom based on the 5 Principles may not understand what they're seeing, especially when Chaos is at the forefront. Think about how to prepare your principal early in the year for how your classroom will be different. What can you share with the administration? How can you let your principal know what he or she might see and how to understand it? Short of buying them a copy of this book, how can you help principals see the value in the 5 Principles?

5. Likewise, parents will find the approach unusual and may respond with confusion or concern. Craft a message for your class website or for back to school night to lay out the Principles for parents and help them understand how your classroom can create better problem solvers. What messages are most important for parents? What can you share with them to help them advance the principles with their children at home?

7 Celebration

Mistakes lead to good places so if you make a mistake take it as a step up the learning ladder.

—Jessica, Age 9, A Very Wise Student

It was incorrectly reported last Friday that today is T-shirt Appreciation Day. In fact, it is actually Teacher Appreciation Day.

—Daily Vidette, Student
Newspaper at Illinois State University

TODAY, WE CELEBRATE

My wife and I did not get married in her family church because their parish sanctuary was too small to accommodate all of our guests. So, we went to the next town, where they had a very large church building. Unfortunately, the parish priest, Monsignor Shoemaker, didn't know either of us, so he met with us several times in the weeks prior to the ceremony, discussing our lives, our families, and our plans as a married couple.

On the day of our wedding, Monsignor led us gracefully through the liturgy and the reading of scripture, followed by his homily.

"Today, we celebrate the union of Michele and Gerald," he began. This was followed by a moving and highly personal sermon on the joys, responsibilities, and beauty of marriage. It was filled with personal details from our meetings that made it clear he not only truly wanted to understand us as a couple, but that he had worked to weave them into his message in a way that made it a moving and emotional tribute to the marriage that was beginning that day. His message truly was one of celebration.

Celebration is just as important in the classroom as it is in a wedding ceremony. We must celebrate our students and their achievements on a daily basis in order to encourage them and to keep them growing. It's a critical part of maintaining a positive, constructive learning environment, and it's directly connected to students' learning and achievement. "People like their workplaces better when time is devoted to celebrating milestones and achievements" (Fisher, Frey, & Pumpian, 2012, p. 139).

Beyond the major accomplishments, though, celebration needs to be an underlying part of the culture of our classrooms. It should permeate every activity, every bit of feedback, and every interaction between people. The strategies in this chapter help you build that culture, but let's first look at the rationale and research behind why celebration is so important.

 ## Validate Effort, Not Answers
MP1, MP3, and MP6

Stanford psychologist Carol Dweck has been doing research for decades on motivation, achievement, and success. Arising out of that research is a deceptively simple, but powerful idea that our mind-set determines our outcome. If we believe that intelligence and ability are fixed characteristics, that we are born with whatever we're ever going to have and our success is a function of that fixed capacity, we attribute our success or failure to that innate ability. The upshot is that we see success or failure as a fixed endpoint. If, on the other hand, we believe that growth is possible, and that we can always learn, change, and improve, then a failure simply becomes a "not yet."

However, the most powerful part of Dweck's research is that mind-set itself is malleable. Even people who have strongly fixed mind-sets can change to a growth mind-set with the right encouragement and feedback.

Dweck points out how we can affect a student's mind-set about learning through the language we use to reinforce his or her work. When we praise students for how smart they are and how good their answers are, they develop a fixed mind-set that believes intelligence is an inherent quality that they either possess or don't, and that learning is something that they either get or don't.

When we validate effort instead, then students can develop a growth mind-set, focusing on steady improvement, and believing that everyone can learn, it just takes work to get there.

> "Being right keeps you in place. Being wrong forces you to explore." (Steven Johnson)

It is easy during math instruction to focus on getting right answers, and to primarily validate student accuracy. Instead, try to focus on recognizing the effort that students put into working through a solution. Any effort that is productive, even if it doesn't ultimately result in an answer, should be rewarded.

Effort that is not productive should be honored and redirected. As Steven Johnson (2010) says, "Being right keeps you in place. Being wrong forces you to explore" (p. 137). For example, if you have a group who have worked hard, but are going down an apparently dead-end route, try something like this: I really appreciate the work you have done on this. You did great work in finding the information you needed to solve the problem, and you are really challenging each other to think deeply about the problem. Do you think you are closer to a solution than you were ten minutes ago? Where do you think you went wrong? Let's back up to this point and try a different tack.

In the last chapter, we briefly touched on the idea of the *adjacent possible*. This is theoretical biologist Stuart Kauffman's (2003) idea that biological organisms are always expanding into functional areas that they don't currently inhabit, but which have closely related (adjacent) and untapped potential (possible). Johnson (2010) believes this concept applies to all areas of innovation, not just biological innovation. If we think of our job as facilitating students' ability to expand their own "adjacent possible," then we start to understand the value of celebration in creating a culture where that happens.

It's critical to remember that "all of us live inside our own private versions of the adjacent possible" (Johnson, 2010, p. 40). What is mundane and ordinary to me may well be a major innovation from my student's point of view. Celebrate the discovery!

Yet, also celebrate the mistakes and the failures. These can generate innovative thinking in the right environment. "Being wrong on its own doesn't unlock new doors in the adjacent possible, but it does force us to look for them" (Johnson, 2010, p. 138). Errors in math create learning and brain growth; correct answers do not (Moser et al., 2011). When students find a solution that does not work, instead of merely telling them the solution is wrong, celebrate the knowledge that they now know one more way that doesn't work, and no longer need to continue to explore that part of the adjacent possible.

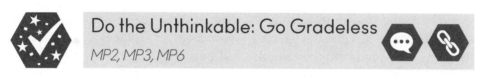

Do the Unthinkable: Go Gradeless
MP2, MP3, MP6

Before we dive into specific celebration-related strategies, I need to address one practice in schools that tends to undermine a culture of celebration: grades. Grading is such a pervasive part of our school routine that it doesn't even occur to most of us that we can do without this practice.

There is a growing movement to do away with grades in classrooms and schools across the United States. In his book, *Assessment 3.0*, Mark Barnes advocates for replacing grades with meaningful feedback. "The GPA is quite possibly the most dangerous tool in education," says the former high school English teacher. "It places a subjective label on students that says very little about academic achievement" (Barnes, 2015, p. 8).

Barnes developed a formula for constructing feedback called **SE2R** which is a perfect fit for the modern mathematics classroom, and supports a culture of Celebration:

- **Summarize**: Provide a one or two sentence summary of what students have accomplished on an activity or project.
- **Explain**: Share detailed observations of what skills or concepts have been mastered based on the specific activity guidelines.
- **Redirect**: Indicate for students the lessons, presentations, or models that need to be reviewed in order to achieve understanding of concepts and mastery of skills.
- **Resubmit**: Encourage students to rework activities, after revisiting prior learning, and resubmit them for further feedback. (Barnes, 2015, p. 34)

This chapter is about the fifth principle of Celebration and celebrating each individual student and his or her learning instead of celebrating a few "winners." It's also about celebrating opportunities for growth, which is where the SE2R formula is so powerful. The emphasis on resubmitting for further feedback takes the emphasis off performance and shifts it to learning. As you will see, all of the Celebration strategies fit into this formula, culminating in a full cycle of Spiral Feedback, which is described in the section for Grades 8–12.

So, why eliminate grades? Because despite attempts to couch grading systems as objective, in truth nearly everything about grading is arbitrary and punitive. From the number of points, to what "counts" toward a grade, to how we award partial credit, the teacher holds all of the power. The student's role degenerates into chasing the grade instead of pursuing deeper understanding.

If you decide to try a no-grades classroom, I encourage you to do the following: read *Assessment 3.0* by Mark Barnes to learn more; follow hashtag #TTOG on Twitter; and join the Teachers Throwing Out Grades groups on Facebook (https://www.facebook.com/groups/teachers throwingoutgrades) and LinkedIn (https://www.linkedin.com/groups /Teachers-Throwing-Out-Grades-TTOG-8239433) to learn about the growing network of teachers who are dedicated to this cause.

CAUTIONS ABOUT CELEBRATION

Finally, a brief caution and an explanation of what Celebration is not. I'm not talking about the type of celebration that pervades youth sports culture where everybody gets a trophy. I'm not talking about meaning-less and trite pats on the head for completing every routine task. And I'm absolutely not talking about ignoring flaws and mistakes and pre-tending that everything's always flowers and rainbows. Errors need to be corrected, and the classroom need not always be fun to be an effective learning environment.

But, we do need to make the classroom a place where it is *safe* to make a mistake, and *safe* to correct those mistakes, and *safe* to have fun or be seri-ous at the appropriate times. In every flaw there is something to be gained, and that growth is a cause for celebration. When we stress the learning instead of the failing, then failures are less scary.

Celebrations in the math classroom must be meaningful and authentic, emerging from a true desire for students to learn and grow. They must also be focused on students' work, process, and effort, not on the end result. This is also an out-growth of Dweck's work:

> In every flaw there is something to be gained, and that growth is a cause for celebration. When we stress the learning instead of the failing, then failures are less scary.

When I taught a group of elementary and middle school children who displayed helpless behavior in school that a lack of effort (rather than lack of ability) led to their mistakes on math prob-lems, the kids learned to keep trying when the problems got tough. They also solved many more problems even in the face of difficulty.

Another group of helpless children who were simply rewarded for their success on easier problems did not improve their ability to solve hard math problems. (Dweck, 2015, para. 7)

Angela Stockman, former English teacher and founder of the WNY Young Writer's Studio, prefers the word "exhibition" rather than celebration, to avoid meaningless praise. Nevertheless, she describes four shifts to make when planning classroom events that help keep them meaningful, and which we should keep in mind when creating a culture of celebration. Although I've modified her descriptions somewhat to adapt to the common vocabulary of the 5 Principles, the core ideas remain. (The following is adapted from Stockman, 2015.)

Focus on Learning. In a celebration event, we share a victory, we read a bit of our work, and we cheer one another on. When we exhibit work in a culture of celebration, we share our thinking, we make it visible to an audience, and we inspire curiosity. The purpose is learning rather than validating. Stockman says students who participate in exhibitions report higher levels of pride and an eagerness to revisit their work in order to make it even better.

Showcase Process and Skill. Celebration events tend to place value on the products that students create. When they exhibit in a more meaningful way, they showcase specific processes or skills. They share in order to reflect, debrief, and help others learn, often from imperfect works in progress. When students share their growing expertise, they begin turning to one another for support rather than relying exclusively on the teacher.

Underscore Perseverance. Celebration events often cover up the trouble, struggle, and uncertainty that students persevered through on their way to finished products. Exhibition in a safe culture of celebration does not deny these realities, but instead illuminates and normalizes them. All students begin to recognize struggle as a part of the problem-solving experience rather than an indication that they are failing to become good problem solvers. Once kids become comfortable with struggle, they're often more willing to take on even greater challenges.

Make It Interactive. Celebration events allow audiences to be a bit more passive than exhibition does. We want students to wonder about, question, and experiment with the approaches that others share. Exhibition embedded in a positive, encouraging culture makes time and space for this important work, and it enables everyone—students and teachers alike—to reap the rewards of powerful assessment practices.

At the beginning of the chapter, I shared a lovely story about my wedding. But, it didn't end there.

About a year after our ceremony, Michele and I happened to attend another wedding in the same church, also officiated by Monsignor Shoemaker. It was just as beautiful as ours, and when we got to the sermon, we expected to hear a similarly personal and touching celebration of two lives. We were partly correct: It was definitely similar. In fact, the homily was word-for-word identical to the one the priest gave at our wedding, except, of course, for the names of the couple and a few minor details.

The realization that the apparently individual message was actually a perfunctory, fill-in-the-blank speech tainted the memory of our special day. If your celebrations for students are similarly perfunctory, kids quickly catch on, and instead of creating a positive and supportive culture, you undermine their willingness to participate.

GRADES K-3

 ### Celebrate Small Victories
MP1

Since we are shifting our mind-set away from only complete and correct answers to questions, it's important that we also honor student efforts toward solving a problem. This is particularly helpful in the early grades, since young children tend to perceive things as right or wrong, and anything that's wrong can't have value.

So celebrate little victories: if a group is working on a tough problem and they have a realization that takes them in a new direction, don't just tell them to keep going, share the win with the class.

Don't just give empty praise, however. Make it meaningful and specific. Focus not just on the success, but how it was successful. "We need to pick up the little things that children do and tell them why that worked," says Wilson McCaskill, expert in student behavior education. "Get them to understand it, perhaps even ask them themselves why it worked and then acknowledge that so they can see how that can lead to the next bigger thing that might work. Here's a thought for you: you should treat praise as you would food; no child should starve from a lack of it just as no child should become obese from too much" (McCaskill, 2010).

> Train yourself to look throughout the day for successes worthy of celebration, and vary the way you recognize them.

Train yourself to look throughout the day for successes worthy of celebration, and vary the way you recognize them. Sometimes, a high-five is sufficient, sometimes it's worth stopping to share with everyone, and sometimes you need to publish a photograph in the class blog.

Teach the students how to celebrate their own work and each other's as well. Lisa Nielsen, a public school educator, administrator, and author, shares a strategy she has used to make this kind of celebration meaningful and powerful:

> We did this by asking students what they were most proud of in their work. At the celebration, they'd read a passage that demonstrated what they were proud of or would explain what they were proud of. I made certificates for them for the achievement they identified.
>
> They loved being noticed for what they wanted to accomplish. Their parents did too. It was a lot of fun and quite special for the kids and parents. The students received one copy of what they published. Another copy went into the school or classroom library. (Teachers Throwing Out Grades, 2015)

GRADES 2-5

 **Daily Math Edit**
MP3, MP6, and MP8

Think about how we build writing fluency and grammar skills in language arts:

1. Explicitly teach important skills and rules.

2. Provide students with "mentor texts" and many examples of excellent prose to read and analyze; students who read a lot also write better.

3. Ask students to review and correct examples which contain errors; this is sometimes done through a "daily edit" routine.

In math, most of our instruction already focuses on the explicit teaching of skills and rules. Mathematical mentor texts are also used, though not as frequently, when we model solutions for students and share examples of worked problems.

When we try to teach students to correct errors in mathematics, however, we tend to focus on simple, careless mistakes. This is inadequate. We should give students more opportunities to analyze and correct mathematical examples with errors. This accomplishes three things:

- It creates a classroom environment where mistakes are normal, and students are empowered to correct those mistakes.
- It teaches students to be more self-reliant in noticing and correcting errors.
- It shows students that errors and mistakes are not the end of the process, but are simply new problems to be solved.

When students get comfortable looking for errors in mathematical work, they get better, and therefore more fluent, at noticing those errors.

You have three sources to draw from in generating a daily math edit problem: teacher created exercises, student work, and prepared examples from outside sources.

You can produce an example yourself by solving a problem and introducing one or more errors into the work. Though time consuming, this option is straightforward and allows you to use examples based on problems that students are already familiar with.

You can also draw from student solutions to previous problems. Before you elect this approach, however, be sure that your classroom culture is developed enough to support this, and always get the student's permission before sharing his or her work. Even if you mask the name and rewrite the solution in your own handwriting, the student is likely to recognize his or her work, and other students may figure it out.

There are few sources of prepared examples for math, but Michael Pershan's *Math Mistakes* blog (http://mathmistakes.org) is excellent. In it, the New York City math teacher collects student work and posts it periodically along with a blog post that discusses the errors in it. The posts are categorized by content area and by Common Core standard, so you should be able to easily find specific examples relevant to a skill or concept you want to reinforce.

As you collect and generate examples, be sure you are including all of these different types of errors:

- **Careless errors**. These are mistakes in copying or computation that result from a lack of precision.
- **Computation errors**. These are mistakes caused by a faulty algorithm, or from taking an inappropriate shortcut.
- **Procedural errors**. These are created when a student chooses the wrong process, such as subtracting instead of adding, dropping a remainder, or finding the volume of a figure instead of the surface area.

- **Errors in reasoning**. This comes from faulty logic and mistaken assumptions during the solution of a problem.
- **Errors in interpreting a solution**. Sometimes students can flawlessly produce a correct solution to a problem, but misinterpret what it means or how it applies. While this is a type of error in reasoning, I list it separately because it's important that students know to look out for it.

GRADES 4–7

 Catch Me If You Can
MP2, MP3, and MP6

I walked into room 108 in the Forum with a mixture of trepidation and excitement. It was just before 8:00 a.m. on the first day of my freshman year at Penn State, and I found a place in the 400–seat lecture hall about two-thirds of the way back, near the left-hand aisle. This was CHEM 12, the introductory chemistry course. The professor began with an overview of the syllabus and general course requirements, then made a rather unusual announcement.

"During every class, I'm going to make one deliberate mistake," he announced. Silence. Either the class was stunned by his revelation, or no one was yet awake. Regardless, he went on. "It's your job to catch my mistake. The first person to notice and correct my mistake will get a candy bar. If the same person manages to catch a second mistake sometime during the semester, you'll get this really nice ballpoint pen."

Though the professor was clearly uninformed about what kinds of prizes would motivate a first-year college student, his concept was unique. I had never experienced a classroom where it was not only *acceptable* to correct your instructor, but *expected*.

And he lived up to his promise. Every lecture contained a deliberate error. Sometimes it was an obvious factual mistake. Sometimes it was more subtle. A few times, it may not have been intentional; it was hard to tell.

However, that was the point. For many of us in the class, the awareness that there would be mistakes in the lecture did several things. First, it made our own preparation more important. Second, it meant we were on higher alert throughout the class to watch for things that didn't make sense or were confusing. And third, it gave us an open invitation—an expectation, really—to challenge and question the professor. This was huge; after all, we were eighteen-year-old freshmen, and he was a college professor. Surely he knew everything and it was our job just to sit at his feet at the crack of dawn three times a week and absorb the immense knowledge and wisdom he was dispensing.

The professor changed these rules. He communicated to us, albeit indirectly, that our job was not to *receive* knowledge, it was to *create* it by understanding concepts deeply enough that we could not only reproduce them on an exam, but we could recognize when they were used incorrectly and fix those mistakes.

I didn't recognize it at the time, but his approach did something else for me. The classroom culture, even in that huge lecture hall, allowed mistakes, and encouraged us to face them directly instead of avoiding them at all costs.

> The classroom culture, even in that huge lecture hall, allowed mistakes, and encouraged us to face them directly instead of avoiding them at all costs.

I want to be very clear here: we were not celebrating the mistake. Errors in themselves aren't desirable: they're still errors. Nevertheless, understanding that mistakes happen, that they are a normal part of the learning and problem-solving process, and that we can and should move forward after them is advantageous. We celebrate acknowledgement of errors, correction of errors, and growth past them.

This strategy works well in the upper elementary grades. Students at this age possess a fragile confidence. They are beginning to recognize their own abilities and talents, and they are also acutely aware of their own weaknesses. Build their confidence by letting them call you out on your mistakes; students love to catch their teacher messing up, even when it's an intentional blunder, and it gives you an opportunity to model that the mistake doesn't undermine your own ability to teach.

A few things to keep in mind as you work this approach into your instruction in math:

- **Plan specific mistakes on a consistent and regular basis.** In a class that meets daily, an intentional mistake a day might be a bit much, but two to three times a week is manageable, and it should happen frequently enough to make it a part of the routine.

- **Plan your mistakes around a range of content and skills.** You want students to learn to recognize a variety of logical and factual mistakes, related to both the current content and concepts they should already have mastered. Besides, if your mistakes are consistently the same type, you become predictable and the strategy loses its power. Several weeks into the semester, I noticed that my chemistry professor almost always timed his mistakes around the midpoint of the lecture, and they were almost always straightforward factual errors. It became easy to guess his routine, and so I stopped looking for other kinds of errors at other times.

- **Know the material you are teaching very well**. I have unfortunately witnessed classrooms where the teacher didn't understand the content well and the instruction was riddled with mistakes. The value of the intentional mistakes as well as the positive culture of accountability evaporates and you lose credibility when errors happen this frequently and unintentionally.

GRADES 6–9

 ## I Guarantee It
MP3, MP5, and MP6

In the middle school years, continue a culture of Celebration in order to build student confidence in their knowledge and abilities to solve problems. This next strategy helps create a space for that confidence, while also making it safe when students are less sure of themselves.

Math teachers commonly have students present their solutions on the board. If the solutions are incorrect, we draw attention to the flaws, correct them, and move on. This can have a detrimental effect on student motivation. Instead of merely attending to accuracy, ask students how sure they are of their work.

When students share their work, in addition to the solution, have them give it a *certainty rating*:

Certainty Rating	Name of the Rating	Definition of the Rating
1	*I have a good start*	I have some ideas that I know are good, but it doesn't all make sense yet
2	*I'm confident*	This is a solid solution, even though there are some places where I'm a little fuzzy
3	*I guarantee it!*	I'm absolutely certain that this answer is correct and complete
?	*I'm not ready yet*	If you like, and if you find it helpful to your students, you can add this fourth rating of a "?," which can be used when students still need more time to work, or when they are completely lost and need help

While using this strategy, it's important to accept all ratings and honor any levels of completeness and certainty. If students frequently rate their work as "1," and the only feedback you give them is "here's what you did wrong," the rating just becomes a substitute for the helpless feeling that students get when they post incorrect answers on the board. If, on the other hand depending on the rating, your class shifts its response then the sharing itself becomes a powerful learning opportunity.

For example, responses to shared work that is rated "1" should emphasize follow-up questions. "What parts make sense to you? Where are you most confused? What kind of help do you think you need?" The discussion then revolves around supporting and celebrating the progress so far.

When a student rates his work as a "2," shift to a balance of positive feedback and supporting guidance. Fellow students should point out where they agree and disagree, and provide suggestions for the places where the solver says they are unclear.

Ratings of "3" indicate the student is extremely confident, and a student with this self-assurance can stand up to more rigorous scrutiny. An appropriate response here is to say, "Prove it!" You can also challenge the student in more depth and ask for evidence to back up the confidence.

With this strategy, a key element of celebration is to recognize students when they grow in confidence. Students who frequently rated their work with a "?" and now begin to use a "1" more often should be honored. Give high fives and cheers.

Patterns in these ratings can also tell you a great deal about the kind of feedback to give and how to help students. You may notice, for example, that certain students often have very solid solutions, but rate them as a "1." Point out their strengths and build their confidence. When students make more frequent errors, your response should be different for students who recognize their errors and designate low ratings to their solutions than for students who express confidence in their incorrect work. The first student example is likely in a position to receive guidance and coaching to assist in building their skills. The second type of student, however, probably needs a deeper analysis and intervention to determine the disconnect and how you can help their growth.

A worthy goal is for students to be able to rate every solution at least a "2," and this helps you both diagnose weaknesses and help students monitor their own growth and learning.

GRADES 8–12

Spiral Feedback
MP2, MP3, MP6, and MP8

At best, I was a mediocre student in high school. Although more than a little of this condition was a result of lousy work habits—fine, call it laziness if you must—I attribute at least some of my performance to poor feedback from my teachers. In elementary school, and for much of middle school, success for me didn't take much effort, and most of the feedback I got from teachers came in the form of praise for my innate abilities. "You're a really good writer," or "You are so smart at math." I didn't feel these assurances internally. I struggled with English, and although looking back I did have a strong talent for math, I usually felt as though I was an impostor among born geniuses in my math classes.

> When I struggled, I attributed it to having reached the limit of my ability instead of having reached a gap in my learning. And no teacher ever told me otherwise.

In retrospect, and knowing what I know now about learning and feedback, I think part of my problem was that I never really knew what my particular strengths and weaknesses were *within* the domain. Although I was "good" at writing and math, I didn't know which parts I knew how to do well and which I needed to work harder at learning. When I struggled, I attributed it to having reached the limit of my ability instead of having reached a gap in my learning. And no teacher ever told me otherwise.

There were also times when I felt as if something should be easy and it wasn't. Others in the class seemed to understand it with no difficulty, and sometimes the teacher would indicate that the concept or problem was a simple one. Instead of asking for help, I suffered in silence rather than admitting that I couldn't handle something that everyone else knew was easy.

I believe if I had received more appropriate, targeted feedback about my work and learning, I would have gained more confidence. Research supports this as well.

Unlike what I received, I call my approach *spiral feedback*, since it is both *balanced* and *cyclical*. To see how this concept works, let's first look at Brookhart's (2008) four dimensions of feedback:

- Timing
- Amount
- Mode
- Audience

Timing

Timing is about when feedback is given and received. To be **balanced**, the feedback timing has to occur within a small window during and after the learning event. Students must be able to connect the feedback with the learning that took place. When feedback occurs too early, it's similar to a GPS navigation system warning you of upcoming turns. Students stop attending to their own thinking and rely on the turn-by-turn directions to coast to a solution. When it occurs too late, students lose the context and don't use the feedback to make valuable adjustments to their thinking.

Cyclical feedback recurs frequently through any learning event. It is not enough to provide feedback once about a specific performance. Come back at least two or three times to reinforce the feedback that's given. Feedback does not all need to come from the teacher. High school students can and should give feedback to each other, and can even learn to give themselves feedback by reviewing their work with a critical eye.

Amount

Feedback must be focused and specific in order to be effective. **Balanced** amounts provide just the right motivation and support for students to progress. If you provide too much feedback, such as marking every flaw in every math problem, students get overloaded and can't process any of it. Too little, on the other hand, and students make little growth or progress.

Hattie (2012) also reports that we should emphasize a type of feedback he calls "disconfirmation." This is feedback that corrects misconceptions, assumptions, and mistaken expectations. In a culture of Celebration, it appears that we should emphasize the wins with lots of praise. Although praise has its place, it needs to be separated from constructive feedback, and we need to also celebrate the mistakes as opportunities to improve. Suppose a student gets part way through the solution to a problem and stops. Good feedback to this student might be, "You seem to have gotten lost in the middle of solving this problem. Can you identify a spot in the process where you are sure of what you were doing? What could you do to get back on track?"

The **cyclical** nature of spiral feedback means that you would come back to this student after the next attempt and give incremental feedback about the new work. Celebrate the growth (I see that you got past the place where you were stuck last time and made it much further) and note the next area to work on (The work you showed here is confusing to me and I'm not clear on how it supports your argument). Focusing on one area at a time, yet having a short turnaround cycle is a simple way to manage the amount of feedback you are giving.

Mode

Feedback comes in three basic modes: verbal, written, and demonstration. In practice, the **balance** comes from using an appropriate blend of modes. Even within one learning situation, it is helpful to combine modes. On a written assignment, give some written feedback, but follow it up with a verbal conversation with the student. Various modes also allow you to **cycle** feedback without it becoming repetitive. For example, you might demonstrate to a student an improved technique for using a protractor, then write a comment on that student's next project pointing out how the technique helped get a more precise solution.

Audience

Audience is perhaps the most important dimension to spiral feedback. You must know your students and tailor your feedback to their needs and where they are in the learning process.

Feedback to a student who is highly successful should be focused on drawing attention to the fine details that can improve performance from good to expert. A student who is having trouble with more fundamental concepts, on the other hand, needs feedback that helps build confidence and basic understanding.

Posamentier, Germain-Williams, and Jaye (2013) point out that "students often do not know what they don't know. . . . Students can become aware of their strengths and weaknesses as learners, and they can learn to take greater control over their own academic performance" (p. 17). You can increase this skill by providing students with targeted feedback, by teaching them to ask their own questions, and to seek feedback when they need it. Understand how specific kinds of feedback can create or dismantle a growth mind-set, however. When students believe their difficulties are a result of innate abilities, they are less likely to ask questions. My own experience as a student confirms this. Avoid comments like: "This is a simple problem; you should have no difficulty with it," or "Why don't you get this? I know you did it last year."

Balance and **cycle** of spiral feedback can assist here. If we know that learning is a process, not an event, and different learners grow and succeed at different rates and depths, we know to approach each student as an individual instead of treating all students as a homogeneous collective.

> If we know that learning is a process, not an event, and different learners grow and succeed at different rates and depths, we know to approach each student as an individual instead of treating all students as a homogeneous collective.

Digital Tools and Resources for Celebration

Notice that I did not include a specific digital tool for each grade level. Many of the tools we discussed previously can be applied to celebration and feedback. Shared documents, wikis, social media, and blogs are all great ways to share and document celebratory feedback and conversations with students. Presentation tools provide opportunities for students to share not just their work, but to celebrate each other's learning.

Two digital tools that allow for specific feedback and individualized celebrations are Kaizena, and digital badges.

Kaizena

Kaizena (https://kaizena.com) is an add-on to Google Docs that allows teachers multiple communication options: to provide feedback to students, to track evidence that they learned specific skills and concepts, and to initiate dialogue with students about their work.

Teachers using Kaizena can attach feedback in the form of text or voice comments directly to student work in Google Drive. The feedback can be associated with a specific highlighted passage, or it can be more general feedback about the document as a whole.

You can also embed links to outside resources, such as instructional videos or reference material related to your comment.

If you are using a rubric to assess specific criteria, characteristics, or skills, you can provide direct feedback about those as well, through the Skills feature. When you highlight a section of the document, you can rate individual skills and provide a related comment explaining your reasoning.

Students can also respond to your feedback directly within the Kaizena interface. When they make revisions to their work, you can pull a new version of the document into Kaizena and continue the spiral feedback process.

Digital Badges

Teachers can create **digital badges** using a variety of online tools. While badges are sometimes built into other educational tools and sites, such as Edmodo (http://www.edmodo.com), teachers can also create custom badge systems in one of the following sites.

- **ClassBadges** (http://www.classbadges.com). This is a basic badge management system. You can create badges for anything you like. Use stock graphics or upload your own. The site simply records your badges and lets you manage your roster to issue badges to your students and maintain records of who has which badges. Students can log in to see the badges they earned.

- **ForAllRubrics** (http://www.forallrubrics.com). This provides a system for managing badges that can be automatically issued when students complete predefined criteria. You can define a rubric for a project. When students meet minimum proficiency levels, the site automatically issues the student a badge. Likewise, you can create a checklist of activities or requirements that will issue a badge on completion. You can also create and issue individual badges not associated with a rubric or checklist. Students can log in to view their badges, check their progress toward incomplete badges, and issue peer-assessment badges to fellow students.

- **Credly** (http://credly.com). Here, you can create badges that aren't connected with any particular group or class. You can explicitly award badges to individuals or groups, or students can claim badges on their own, using a claim code you provide, or by submitting their own evidence to the site. You can then automatically award the badge based on the code, or you can review the evidence first before granting the badge.

All of these sites generate badges that are compatible with Mozilla's **Open Badges** system (http://openbadges.org), making them portable and universal for other badging systems.

In addition to creating badges for academic accomplishments, consider badges for simple things, such as "Great Problem-Solving Insight" or "Cool Collaborator" or "Caught the Teacher Making a Mistake." Some badges should be simple to earn, and others should be more complex. You can even create badges for students to award each other, or ask students to invent their own badges to earn and award.

The real power here is that badges don't have to be of equal value or meaning, and students can build their own unique collection of badges to celebrate all of their individual accomplishments and achievements.

TECHNOLOGY INTEGRATION FOR CELEBRATION: CASE STUDY

Kaizena

Ruth Eichholtz is a math teacher at The York School in Toronto, Ontario. She uses Kaizena with her students to support Celebration and engage them in spiral feedback.

Our students in grades 11 and 12 are required to write papers exploring mathematical concepts, and relate those concepts to areas of personal interest. This is not a straightforward task; the process begins with topic selection, and includes research, calculation, analysis and interpretation. Used as a form of summative assessment, these projects represent a significant learning experience for high school mathematicians.

Topic selection and outline planning are scheduled, informal checkpoints where the students receive verbal and/or written feedback on process. They then perform individual research or data collection, followed by their own calculations and analysis. Students submit PDF copies of their drafts to their Google Drive folder, where I can directly access and open their files in Kaizena.

In Kaizena, I use audio recording to give each individual student praise, suggestions, and corrections directly related to the assessment criteria. The audio feedback has proven very popular with students, who report that it was effective in helping with revision. In a recent survey, I found that most students listened to the recorded comments at least twice, often more. This allowed them sufficient time and repetition to interpret and apply teacher feedback in finalizing their work. In addition, most students indicated that they prefer verbal to written feedback on this project. Given the preference for verbal feedback, audio recording is highly efficient and reliable, as the comments are never lost or forgotten. It has become an integral part of the learning process for this task.

FOR PLC AND STUDY GROUPS

1. The major theme of this chapter is turning flaws and mistakes into growth opportunities. Instead of penalizing students with points off for an error, give them feedback and a chance to learn, then celebrate when they succeed. Identify the practices in your classroom and school that already support this approach. Come up with a plan for sharing and reinforcing these. What opportunities does this provide for talking more in depth to stakeholders about the 5 Principles?

(Continued)

(Continued)

2. What practices in your school or district tend to undermine a culture of Celebration? For example, do you have an honor roll that is based solely on GPA? Or do your morning announcements trumpet the achievements of the sports teams, but not the participants in the math competition? Outline steps you can take to shift the celebrations in your school into ones that support a classroom culture based on the 5 Principles. What actions can you take to promote a culture of Celebration in your school? What are the risks?

3. A growing area of research is the idea of promoting "grit" in students, defined by psychologist Angela Duckworth as perseverance and passion for long-term goals (Duckworth et al, 2007). What aspects of grit are compatible with a culture of Celebration? Where do they seem to conflict? How can you reconcile the two?

4. Spiral feedback can be an inefficient way to support students. Done well, it is individualized and frequent, and it is detailed enough to give meaningful information to the learner. What are some ways you can provide effective spiral feedback without becoming bogged down in long hours of writing comments to students? How do the concepts of *balance* and *cycle* help you give feedback without it becoming unmanageable? How can technology help you improve the feedback you give to students without giving up valuable time needed for other aspects of your job?

5. Now that you have studied all 5 Principles, identify some "low-hanging fruit" in your practice that might be ripe for exploration. What is one thing you could change tomorrow that could make the greatest impact on student learning? Remembering that this change is meant as a starting point, not an end goal, where are the obvious leverage points that can start moving you most easily toward a classroom culture of innovative thinking?

8 Becoming a Problem-Solving Classroom

Unless you try to do something beyond what you have already mastered, you will never grow.

—Ronald E. Osborn,
Distinguished Teacher and Writer

Changing the way we teach is a lot easier than changing the way students learn.

—Robert John Meehan,
The Voice of the American Teacher

TRANSFORMATION IS POSSIBLE

You may feel after getting this far into this book that your classroom is in a good place. You may already use many of these strategies, or similar ones, and you may simply need some minor adjustments and tweaking to become a classroom fully founded on the 5 Principles.

> The shift is not going to happen merely because you read and understand this book.

On the other hand, you may realize that your practices are extremely traditional and that you have some work to do to revamp your classroom to support a different kind of culture.

Either way, the shift is not going to happen merely because you read and understand this book. It's going to take self-assessment, planning, and maintenance. No change is simple, and for some, the shift to a modern mathematics classroom and a culture of innovative thinking is going to take significant effort and time.

In this chapter, we explore a few ways to tackle this process.

SELF-ASSESSMENT

To begin, you need to assess your classroom's existing culture. Take some time to reflect on your own practices and build a profile of your current math classroom. An ideal time to do this is immediately after the end of a school year, when things are fresh in your mind and you're beginning the transition to planning for the new year. But, if you're reading this book in the middle of a school year and want to start making adjustments now, I encourage you to begin immediately.

First, take an inventory of your own practices. Brainstorm the routines, procedures, habits, norms, strategies, techniques, and expectations you use. Write each on an index card or a sticky note. Then sort them according to which Principle or Principles they most closely align with. You should end up with a list of strategies for each of the 5 Principles.

Next, build a 3–column chart for each of the Principles (see Figure 8.1).

Figure 8.1   Chart for Categorizing Practices After Sorting by Principle
Create a copy of this chart for each of the 5 Principles.

Promotes	Unsure/Both	Erodes

Reflect on each strategy in your list, and think about whether the strategy promotes the culture of innovation and problem solving or erodes it. If you aren't sure, or if it may do some of both, list it in the middle column.

Strategies that promote the culture are candidates to keep, while those which erode are candidates to eliminate or revise. The ones in the middle you can deal with later.

Focus at this time on one of the 5 Principles which you most want to improve. You might choose the one which is already strongest so that you can solidify that area, or you might choose the one where you believe you have the most work to do. Think about what fits your individual needs and those of your students.

Next, it is helpful to directly observe the strategies in your classroom. You can do this either by recording yourself teaching, or by inviting a trusted colleague into your classroom to conduct the observation for you. If you have a strong, collaborative relationship with your administrator, he or she can help here too, but be sure that you establish up front that you are looking for assistance with growth and not a formal evaluation.

You, or the observer, should now monitor at least one complete lesson, although examining two or three lessons over the course of several days is preferable. During the observations, record relevant data about what is taking place during the lessons. Following are recommendations about what to look for when focusing on each of the principles.

General Recommendations for All Principles

- What verbal and nonverbal cues does the teacher use that promote or erode a culture of problem solving?
- What verbal and nonverbal responses do students exhibit that indicate that problem solving is an integral part of the classroom culture?
- What daily routines are in place that promote or erode a culture of problem solving?

Conjecture

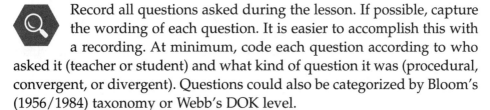 Record all questions asked during the lesson. If possible, capture the wording of each question. It is easier to accomplish this with a recording. At minimum, code each question according to who asked it (teacher or student) and what kind of question it was (procedural, convergent, or divergent). Questions could also be categorized by Bloom's (1956/1984) taxonomy or Webb's DOK level.

Review the collected data:

- What patterns do you see?
- What proportion of questions are asked by the teacher and students?
- What type of question is asked most often by students, and by the teacher?
- How often does the teacher respond to student questions with another probing question?

Other things you might gather data about:

- How often do students provide evidence to support their assertions?
- How often does the teacher ask for evidence?
- What kinds of metacognitive questions are asked and how often?

Communication

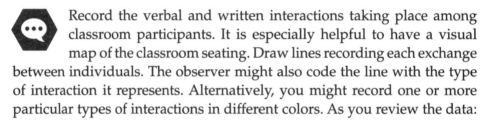 Record the verbal and written interactions taking place among classroom participants. It is especially helpful to have a visual map of the classroom seating. Draw lines recording each exchange between individuals. The observer might also code the line with the type of interaction it represents. Alternatively, you might record one or more particular types of interactions in different colors. As you review the data:

- What patterns do you see?
- Is there a distinct focal point or "hub" around which most of the communication is taking place, or is it distributed around the room?
- Do students interact and communicate broadly among the whole group, or is discussion focused in pairs and smaller groups?
- Do you find a few individuals dominating the conversation?
- How often are students writing for each other?
- How often are students reading each others' writing about math?

Collaboration

 Observe the learning activities taking place and consider the following questions:

- How often do students work alone, in pairs, and in small groups?
- When students are working in pairs and groups, are activities structured in a way that promotes every student learning all of the content, or do specific individuals carry most of the load?
- Do students work together before teacher instruction, after it, or both?

- Is instruction geared toward the whole class, small groups, or individuals? What is the distribution of these types of instruction over time?
- When the teacher is working with a small group or individual, what are the other students doing?

Chaos

 Observe the flow and process of instruction in the classroom and consider the following questions:

- Is there a general tolerance in the classroom for messiness?
- Does the teacher frequently rescue students who are struggling, or are they allowed to work through the confusion?
- When the teacher assists struggling students, does he or she ask questions that lead or questions that provoke thinking?
- Is there a reasonable amount of productive struggle throughout the classroom, or are students floundering?

Celebration

 Observe the general environment and interactions within the classroom. Ask these questions:

- How do students respond to their own errors and to the errors of other students?
- How do students respond to their own successes (large or small) and to the successes of other students?
- How does the teacher respond to student errors and to his or her own errors?
- How does the teacher respond to student successes (large or small)?
- Does the teacher share his or her own successes with the class? Are they framed in terms of growth (I accomplished this through hard work) or a fixed mindset (I'm really good at math)?
- Watch one or two students closely for a period of time. What do these students do or say that indicates either a fixed or a growth mindset?
- What does the teacher do and say that sets a tone conducive to a growth mindset in the classroom?

GOAL-SETTING AND PLANNING

Following this self assessment, develop an action plan to transform your classroom based on the 5 Principles. This transformation will not happen overnight, however, and you should not try to implement everything at once.

I recommend you choose one Principle and work on it first. If one of the Principles speaks to you loudly, either because you are already doing it well and it is very compatible with your present approach to teaching, or because you feel strongly it is the one you need the most work on and want to tackle it first, then choose it. You are far more likely to stick with a plan of change when you are working on something that is personally meaningful to you.

> You are far more likely to stick with a plan of change when you are working on something that is personally meaningful to you.

On the other hand, if none of the Principles leaps out at you, then begin with Conjecture and work your way through in the order presented in this book. Though the 5 Principles are of equal importance, there is logic to the order and your classroom transformation will have a natural flow if you follow it.

Once you've chosen your focus area, revisit the data you collected and develop an action plan. Figure 8.2 can ensure that you are considering all elements of the classroom that contribute to a good environment.

Define one overarching goal for the Principle and a subgoal for each area of focus. For example, if you are working on Conjecture, your

Figure 8.2 Action Planning Form for the Goal Areas

Area of Focus	Action Steps	Time Frame	Person Responsible	Resources Needed
Physical space				
Curriculum materials and resources				
Instructional practices				
Assessment practices				
Daily routines and procedures				
Social/emotional environment				

overarching goal might be that by the end of the first quarter, students will ask at least half of the questions during a typical lesson, and that students will have the confidence to answer each others' questions instead of waiting for the teacher to answer them. Think about how the physical arrangement of the room could support this goal. What changes can you make in the materials you use, or how can you use your materials differently? What daily routines and instructional strategies can you select from those outlined in this book to help you achieve this goal?

Your overall plan should map out when you will work on each of the 5 Principles. Though it is best to work on one Principle at a time, don't make the mistake of thinking you have to master one Principle before you move on to the next. For the same reason that your students should practice math skills in the context of problem solving, you should cycle through all 5 Principles to introduce them into your teaching, then cycle back to reinforce and add more depth. A good rule of thumb is to spend three or four weeks focusing on one Principle before moving on to the next. This allows you to complete approximately two cycles during a school year so that by the end of that year, you have established new habits, and are ready to begin again the following year.

MAINTAINING YOUR MOMENTUM

Changing your classroom is easy. Sustaining that change beyond the first day is difficult.

You are probably familiar with the great deal of research available on what it takes to change a habit, and what it takes to maintain the new habit. Rather than rehashing that information, I want to point you to what influenced the most change in me, and in many ways, what made this book possible in the first place: my professional network.

> Changing your classroom is easy. Sustaining that change beyond the first day is difficult.

Steven Johnson talks about the importance of liquid networks to innovation. Members of a constantly evolving network create exposure to new ideas, and challenges to those ideas, making us smarter. "It's not that the network itself is smart; it's that the individuals get smarter because they're connected to the network" (Johnson, 2010, p. 58).

Networks and the people in them are extremely helpful in supporting and sustaining change. Unless it was selected for a faculty book study, it is very possible that you are the only teacher in your school, or even district, who is reading this book. As a result, finding someone else to check in with about the book and to bounce ideas off of is going to be difficult.

Enter the power of the professional learning network: your **PLN**. If you reach outside your immediate locality, you will find other teachers with like minds and like interests around the world, many of whom are using the 5 Principles in their own classrooms.

Having a partner or group who knows your plan and helps keep you on track is crucial to your success. Where can you begin to build this network if you don't already have one? Standard social media are excellent places to start: there are large, active communities of teachers on Twitter, Facebook, Edmodo, Pinterest, and LinkedIn. You can also connect directly with me by following me on Twitter at @geraldaungst, and there are links to my other online profiles at my website, http://www.geraldaungst.com.

Corwin also has an entire collection of books dedicated to building and growing your PLN. The Connected Educator series (http://www.corwin.com/connectededucators) is written by phenomenal educators, many of whom are part of my own PLN. They have been developing their own networks for years and can guide you in building yours.

> It's my hope that after a year working in a classroom dedicated to the 5 Principles, your students will never again believe that they are not good at math.

Finally, there is a growing community of educators specific to this book. To join the discussion, visit http://www.geraldaungst.com/5cmath.

Though this is the end of the book, I hope it is the beginning of our journey toward the 5 Principles together. You now have a bigger toolbox and a framework for understanding how you can make your students better problem solvers and more innovative thinkers. This book began with a statement heard in many classrooms today: "But, I'm just not that good at math." It's my hope that after a year working in a classroom dedicated to the 5 Principles, your students will never again believe that they are not good at math. You are in a position to launch the world's next Grace Hopper or Steve Jobs. Now, go give them their own toolboxes and the confidence to tackle whatever problems they may face in the future.

Appendix

Digital Tools

In this appendix, I have compiled a list of all of the digital tools and resources mentioned throughout the book. The first section lists them by chapter, the second has the same tools listed alphabetically for easy reference. In addition, this list is available online at http://www.geraldaungst .com/5cmath/tools.

DIGITAL TOOLS BY CHAPTER

Chapter 3. Conjecture

Educreations (http://www.educreations.com)

ShowMe (http://www.showme.com)

PixiClip (http://www.pixiclip.com)

Screencast.com

Screencast-O-Matic.com

MathPickle (http://www.mathpickle.com)

MathPickle Unsolved Problems (http://mathpickle.com/unsolved-k-12)

The National Library of Virtual Manipulatives (http://nlvm.usu.edu)

Data.gov (http://data.gov)

American Statistical Society's useful sites for teachers (http://www.amstat.org/education/usefulsitesforteachers.cfm)

GeoGebra (http://www.geogebra.org)

WeLearnedIt (http://welearned.it)

Chapter 4. Communication

Edmodo (http://www.edmodo.com)

Google Classroom (http://classroom.google.com)

Blogger (http://www.blogger.com)

Wordpress (http://www.wordpress.com)

Kidblog (http://www.kidblog.org)

Edublogs (http://www.edublogs.org)

Piktochart (http://piktochart.com)

Infogr.am (http://infogr.am)

Audacity (http://audacity.sourceforge.net)

GarageBand (https://www.apple.com/mac/garageband)

Creating a Wordpress.com podcast blog
(http://en.support.wordpress.com/audio/podcasting)

Creating a self-hosted Wordpress podcast blog
(http://codex.wordpress.org/Podcasting)

Tony Vincent's introduction to podcasting
(http://www.readingrockets.org/article/creating-podcasts-your-students)

Podcasting Student Broadcasts at Edutopia
(http://www.edutopia.org/podcasting-student-broadcasts)

Creative Commons Podcasting Legal Guide
(http://creativecommons.org/podcasting)

Chapter 5. Collaboration

Skype in the Classroom project (http://education.skype.com)

Mystery Skype game (https://education.skype.com/mysteryskype)

Skype virtual field trips (https://education.skype.com/partners)

Google Drive (http://drive.google.com)

Google Apps for Education
(https://www.google.com/work/apps/education)

Padlet (http://padlet.com)

Padlet for the 5 Principles (http://padlet.com/geraldaungst/5CMath)

Tricider (http://www.tricider.com)

Tricider example, locker problem
(http://www.tricider.com/brainstorming/2m0nmF1Xbet)

Trello (https://trello.com)

Evernote (https://evernote.com)

Model-Eliciting Activities: Case Studies for Kids
(https://engineering.purdue.edu/ENE/Research/SGMM/CASESTUDIES
KIDSWEB/index.htm)

Chapter 6. Chaos

Animoto (http://animoto.com)

All Bets are Off Outside Math Class (http://www.edutopia.org/blog
/all-bets-off-outside-math-class-matt-levinson)

Art of Problem Solving (http://artofproblemsolving.com)

Bedtime Math (http://bedtimemath.org)

Dan Meyer's Three Act Problems (http://threeacts.mrmeyer.com)

The Educational Value of Math Puzzles (http://www.edutopia.org/blog
/recreational-educational-value-math-puzzles-deepak-kulkarni)

MathPickle Unsolved Problems (http://mathpickle.com/unsolved-k-12)

Problems of the Month from Inside Mathematics
(http://www.insidemathematics.org/problems-of-the-month)

POW: Problems of the Week
(http://www.math.com/teachers/POW.html)

Problems of the Week from the Math Forum @ Drexel University
(http://mathforum.org/pows)

Recreational Math (http://www.math.com/teachers/recreational.html)

Stella's Stunners (http://ohiorc.org/for/math/stella)

Minecraft (https://minecraft.net)

Minecraft Firework Rockets
(http://minecraft.gamepedia.com/Firework_Rocket)

World of Warcraft (WoW) (http://us.battle.net/wow/en)

WoW in School: A Hero's Journey (http://wowinschool.pbworks.com)

The Superhero Database (http://www.superherodb.com)

So Who Wins: The Any Character in the World Battles (http://www.sowhowins.com)

Dan Meyer's Makeover Monday (http://blog.mrmeyer.com/category/makeovermonday)

Scratch (http://scratch.mit.edu)

Alice (http://www.alice.org)

PlayNIM in Scratch (https://scratch.mit.edu/projects/16135815)

Inform (http://www.inform7.com)

Chapter 7. Celebration

ClassBadges (http://www.classbadges.com)

ForAllRubrics (http://www.forallrubrics.com)

Credly (http://credly.com)

Mozilla's Open Badges system (http://openbadges.org)

Chapter 8. Becoming a Problem-Solving Classroom

Corwin Connected Educator series (http://www.corwin.com/connectededucators)

5 Principles Discussion Group (http://www.geraldaungst.com/5cmath)

DIGITAL TOOLS LISTED ALPHABETICALLY

5 Principles Discussion Group (http://www.geraldaungst.com/5cmath)

Alice (http://www.alice.org)

American Statistical Society's useful sites for teachers (http://www.amstat.org/education/usefulsitesforteachers.cfm)

Animoto (http://animoto.com)

Art of Problem Solving (http://artofproblemsolving.com)

Audacity (http://audacity.sourceforge.net)

Bedtime Math (http://bedtimemath.org)

Blogger (http://www.blogger.com)

ClassBadges (http://www.classbadges.com)

Corwin Connected Educator series
(http://www.corwin.com/connectededucators)

Creative Commons Podcasting Legal Guide
(http://creativecommons.org/podcasting)

Credly (http://credly.com)

Dan Meyer's Makeover Monday
(http://blog.mrmeyer.com/category/makeovermonday)

Dan Meyer's Three Act Problems (http://threeacts.mrmeyer.com)

Data.gov (http://data.gov)

Edmodo (http://www.edmodo.com)

Edublogs (http://www.edublogs.org)

The Educational Value of Math Puzzles (http://www.edutopia.org/blog
/recreational-educational-value-math-puzzles-deepak-kulkarni)

Educreations (http://www.educreations.com)

Evernote (https://evernote.com)

ForAllRubrics (http://www.forallrubrics.com)

GarageBand (https://www.apple.com/mac/garageband)

GeoGebra (http://www.geogebra.org)

Google Apps for Education
(https://www.google.com/work/apps/education)

Google Classroom (http://classroom.google.com)

Google Drive (http://drive.google.com)

Infogr.am (http://infogr.am)

Inform (http://www.inform7.com)

Kidblog (http://www.kidblog.org)

Makeover Monday, Dan Meyer
(http://blog.mrmeyer.com/category/makeovermonday)

MathPickle (http://www.mathpickle.com)

MathPickle: Unsolved Problems for K–12
(http://www.mathpickle.com/unsolved-k–12)

Minecraft (https://minecraft.net)

Minecraft Firework Rockets
(http://minecraft.gamepedia.com/Firework_Rocket)

Model-Eliciting Activities: Case Studies for Kids
(https://engineering.purdue.edu/ENE/Research/SGMM/CASESTU
DIESKIDSWEB/index.htm)

Mystery Skype game (https://education.skype.com/mysteryskype)

National Library of Virtual Manipulatives (http://nlvm.usu.edu)

NIM game example, PlayNIM in Scratch
(https://scratch.mit.edu/projects/16135815)

Open Badges system by Mozilla (http://openbadges.org)

Padlet (http://padlet.com)

Padlet for the 5 Principles (http://padlet.com/geraldaungst/5CMath)

Piktochart (http://piktochart.com)

PixiClip (http://www.pixiclip.com)

PlayNIM in Scratch (https://scratch.mit.edu/projects/16135815)

Podcasting Legal Guide (http://creativecommons.org/podcasting)

Podcasting Student Broadcasts at Edutopia
(http://www.edutopia.org/podcasting-student-broadcasts)

Podcasting, introduction by Tony Vincent
(http://www.readingrockets.org/article/creating-podcasts-your-students)

Podcasting: Creating a self-hosted Wordpress blog
(http://codex.wordpress.org/Podcasting)

Podcasting: Creating a Wordpress.com blog
(http://en.support.wordpress.com/audio/podcasting)

POW: Problems of the Week (http://www.math.com/teachers/POW.html)

Problems of the Month from Inside Mathematics
(http://www.inside athematics.org/problems-of-the-month)

Problems of the Week from the Math Forum @ Drexel University
(http://mathforum.org/pows)

Recreational Math (http://www.math.com/teachers/recreational.html)

Scratch (http://scratch.mit.edu)

Screencast-O-Matic.com

Screencast.com

ShowMe (http://www.showme.com)

Skype in the Classroom project (http://education.skype.com)

Skype virtual field trips (https://education.skype.com/partners)

So Who Wins: The Any Character in the World Battles
(http://www.sowhowins.com)

Stella's Stunners (http://ohiorc.org/for/math/stella)

Superhero Database (http://www.superherodb.com)

Three Act Problems by Dan Meyer (http://threeacts.mrmeyer.com)

Trello (https://trello.com)

Tricider (http://www.tricider.com)

Tricider example, locker problem
(http://www.tricider.com/brainstorming/2m0nmF1Xbet)

WeLearnedIt (http://welearned.it)

Wordpress (http://www.wordpress.com)

Wordpress: Creating a self-hosted podcast blog
(http://codex.wordpress.org/Podcasting)

Wordpress: Creating a Wordpress.com podcast blog
(http://en.support.wordpress.com/audio/podcasting)

World of Warcraft (WoW) (http://us.battle.net/wow/en)

WoW in School: A Hero's Journey (http://wowinschool.pbworks.com)

References

Barnes, M. (2015). *Assessment 3.0*. Thousand Oaks, CA: Corwin.

Bloom, B. S. (Ed.). (1956/1984). *Taxonomy of educational objectives, Handbook I: The cognitive domain*. New York: Longman.

Boaler, J. (2014, September 10). *The mathematics of hope: Moving from performance to learning in mathematics classrooms* [Web log post]. Retrieved from http://www .heinemann.com/blog/the-mathematics-of-hope-moving-from-performance-to-learning-in-mathematics-classrooms

Bogomolny, A. (2014). Mathematics as a language from *Interactive Mathematics Miscellany and Puzzles*. Retrieved from http://www.cut-the-knot.org/language/index.shtml.

Boykin, A. W., & Noguera, P. (2011). *Creating the opportunity to learn: Moving from research to practice to close the achievement gap*. Alexandria, VA: ASCD.

Brookhart, S. M. (2008). *How to give effective feedback to your students*. Alexandria, VA: ASCD.

Brown, S. I. (1993). Towards a pedagogy of confusion. In A. M. White (Ed.), *Essays in humanistic mathematics* (pp. 107–121). Washington, DC: The Mathematical Association of America.

Chesser, L. (2014, June 26). *50 questions to help students think about what they think* [Web log entry]. Retrieved from http://www.teachthought.com/learning/metacognition-50–questions-help-students-think-think

Claessens, A., Engel, M., & Curran, F. C. (2014). Academic content, student learning, and the persistence of preschool effects. *American Educational Research Journal, 51*(2), 403–434.

Colvin, R. L., & Jacobs, J. (2010, April 7). *Rigor: It's all the rage, but what does it mean?* Retrieved from http://hechingerreport.org/content/rigor-its-all-the-rage-but-what-does-it-mean_2222

Crannell, A. (1994). *A guide to writing in mathematics classes*. Retrieved from https://edisk.fandm.edu/annalisa.crannell/writing_in_math/guide.html

Dean, D., Jr., & Kuhn, D. (2007). Direct instruction vs. discovery: The long view. *Science Education, 91*(3), 384–397.

Duckworth, A. L., Peterson, C., Matthews, M. D., & Kelly, D. R. (2007). Grit: Perseverance and passion for long-term goals. *Journal of Personality and Social Psychology, 92*(6), 1087–1101.

Duncker, K. (1945). On problem solving. *Psychological monographs, 58*(5), American Psychological Association.

Dweck, C. (2006). *Mindset*. New York: Ballantine.

Dweck, C. (2015, January 1). The secret to raising smart kids. *Scientific American Special Editions*, 23(5s). Retrieved from http://www.scientificamerican.com/article/the-secret-to-raising-smart-kids1

Engel, M., Claessens, A., & Finch, M. (2013). Teaching students what they already know? The misalignment between instructional content in mathematics and student knowledge in kindergarten. *Educational Evaluation and Policy Administration, 35,* 157–178.

Exercise (mathematics). (2014, May 9). In *Wikipedia, the free encyclopedia.* Retrieved 01:09, January 25, 2015, from http://en.wikipedia.org/w/index.php?title=Exercise(mathematics)&oldid=607825708

Fisher, D., Frey N., & Pumpian, I. (2012). *How to create a culture of achievement in your school and classroom.* Alexandria, VA: ASCD.

Garner, R. (2015, March 20). Finland schools: Subjects scrapped and replaced with 'topics' as country reforms its education system. *The Independent.* Retrieved from http://www.independent.co.uk

Gavin, M. K., Chapin, S., Dailey, J., & Sheffield, L. (2006). *Unraveling the mystery of the MoLi Stone: Place value and numeration.* Dubuque, IA: Kendall Hunt.

Gelman, A., & Nolan, D. (2002). *Teaching statistics: A bag of tricks.* New York: Oxford University Press.

Getzels, J. W. (1982). The problem of the problem. In R. Hogarth (Ed.), *New directions for methodology of social and behavioral science: Question framing and response consistency* (No. 11). San Francisco, CA: Jossey-Bass.

Gilbert, L. (1981). *Particular passions: Grace Murray Hopper.* Women of Wisdom Series (1st ed.). New York: Author.

Gillespie, L., & Lawson, C. (2010). *WoW in school: A hero's journey.* Retrieved from http://wowinschool.pbworks.com/f/WoWinSchool-A-Heros-Journey.pdf

Ginsburg, D. (2014, March 31). *Engaging math students in productive struggle* [Web log post]. Retrieved from http://blogs.edweek.org/teachers/coach_gs_teaching_tips/2014/03/engaging_math_students_in_productive_struggle.html

Green, E. (2014, July 23). Why do Americans stink at math? *The New York Times.* Retrieved from http://www.nytimes.com/2014/07/27/magazine/why-do-americans-stink-at-math.html

Gruber, M. J., Gelman, B. D., & Ranganath, C. (2014). States of curiosity modulate hippocampus-dependent learning via the dopaminergic circuit. *Neuron, 84*(2), 486–496.

Hamilton, G. (2011, May 28). *Let's abolish elementary mathematics* [Video file]. Retrieved from http://youtu.be/3sN3dEVeMb8

Harvard Business Review. (2000, May 1). *Use and misuse of statistics.* Retrieved from https://hbr.org/product/use-and-misuse-of-statistics/an/U0603C-PDF-ENG

Hattie, J. (2012). *Visible learning for teachers.* New York: Routledge.

Hiebert, J., & Grouws, D. A. (2007). The effects of classroom mathematics teaching on students' learning. In F. K. Lester (Ed.), *Second handbook of research on mathematics teaching and learning* (pp. 371–404). Charlotte, NC: Information Age.

Holtz, G., & Malen, M. L. (n.d.). Jumping frogs. In *Open-ended math problems, April problems.* Philadelphia, PA: The Franklin Institute.

Huff, D. (1982). *How to lie with statistics.* New York: W. W. Norton.

Jackson, R. R., & Lambert, C. (2010). *How to support struggling students.* Alexandria, VA: ASCD.

Johnson, D. W., Johnson, R. T., & Stanne, M. E. (2000). *Cooperative learning methods: A meta-analysis.* Minneapolis, MN: University of Minnesota Press. Retrieved from http://www.tablelearning.com/uploads/File/EXHIBIT-B.pdf

Johnson, S. (2010). *Where good ideas come from* (Kindle Version). New York: Penguin Group.

Juliani, A. J. (2014). *Inquiry and innovation in the classroom: Using 20% time, genius hour, and PBL to drive student success.* New York: Routledge.

Kaplinsky, R. (2015, February 4). *Tool to distinguish between depth of knowledge levels.* Retrieved from http://robertkaplinsky.com/tool-to-distinguish-between-depth-of-knowledge-levels/

Kauffman, S. (2003, November 9). *The adjacent possible.* Retrieved from http://edge.org/conversation/the-adjacent-possible

Kilpatrick, J., Swafford, J., & Findell, B. (2001). *Adding it up: Helping children learn mathematics.* Washington, DC: National Academy Press.

Kimball, M., & Smith, N. (Oct 28, 2013). The myth of I'm bad at math. *The Atlantic.* Retrieved from http://www.theatlantic.com/education/archive/2013/10/the-myth-of-im-bad-at-math/280914/

Klahr, D., & Nigam, M. (2004). The equivalence of learning paths in early science instruction: Effects of direct instruction and discovery learning. *Psychological Science, 15,* 661–667.

Korn, B. (2014). *Responsible thinking: How can we avoid believing things that aren't true?* Retrieved from http://www.truthpizza.org/main.htm

Lesh, R., & English, L. (2001). Paper airplanes in *Case studies for kids.* Retrieved from https://engineering.purdue.edu/ENE/Research/SGMM/CASESTUDIESKIDSWEB/paperairplanes.htm

Lucas, C. G., Gopnik, A., & Griffiths, T. L. (2010). When children are better (or at least more open-minded) learners than adults: Developmental differences in learning the forms of causal relationships. In S. Ohlsson, & R. Catrambone (Eds.), *Proceedings of the 32nd annual conference of the Cognitive Science Society* (pp. 28–52). Austin, TX: Cognitive Science Society.

Lyman, F. (1981). The responsive classroom discussion. In A. S. Anderson (Ed.), *Mainstreaming Digest.* College Park, MD: University of Maryland College of Education.

Malouff, J. (n.d.). *Over fifty problem-solving strategies explained.* Retrieved from http://www.une.edu.au/about-une/academic-schools/bcss/news-and-events/psychology-community-activities/over-fifty-problem-solving-strategies-explained

Marzano, R. J., & Pickering, D. J. (2005). *Building academic vocabulary: Teacher's manual.* Alexandria, VA: Association for Supervision & Curriculum Development.

Mason, J., Burton, L., & Stacey, K. (2010). *Thinking mathematically* (Kindle Version). Upper Saddle River, NJ: Pearson.

McCaskill, W. (2011, May 4). Should we praise our children? *Practical parenting* [Video file, episode 9] Retrieved from https://www.youtube.com/watch?v=WExKlI6XXZ4

Medina, J. (2014). *Brain rules: 12 principles for surviving and thriving at work, home and school* (2nd ed.). Seattle, WA: Pear Press.

Moser, J., Schroder, H. S., Heeter, C., Moran, T. P., & Lee, Y. H. (2011). Mind your errors: Evidence for a neural mechanism linking growth mind-set to adaptive post error adjustments. *Psychological Science, 22,* 1484–1489.

National Aeronautics and Space Administration (NASA). (2015, January 13). *GISS Surface Temperature Analysis*. Retrieved from http://data.giss.nasa.gov/gistemp/graphs_v3/

The National Council of Teachers of Mathematics. (2000). *Principles and standards for school mathematics*. Reston, VA: Author.

National Governors Association Center for Best Practices, and Council of Chief State School Officers [NGA/CCSSO]. (2010). *Common Core State Standards*. Washington, DC: Author.

Newell, A., & Simon, H. A. (1972). *Human problem solving*. Englewood Cliffs, NJ: Prentice Hall.

Nunnery, J. A., Chappell, S., & Arnold, P. (2013). A meta-analysis of a cooperative learning model's effects on student achievement in mathematics. *Cypriot Journal of Educational Sciences, 8*(1), 34–48.

Ostroff, W. (2012). *Understanding how young children learn: Bringing the science of child development to the classroom*. Alexandria, VA: ASCD.

Paulos, J. A. (2001). *Innumeracy: Mathematical illiteracy and its consequences*. New York: Hill and Wang.

Pink, D. (2005). *A whole new mind: Why right-brainers will rule the future*. New York: Riverhead Books.

Posamentier, A. S., Germain-Williams, T. L., & Jaye, D. I. (2013). *What successful math teachers do, grades 6–12: 80 research-based strategies for the Common Core-aligned classroom*. Thousand Oaks, CA: SAGE.

Pretz, J. E., Naples, A. J., & Sternberg, R. J. (2003). Recognizing, defining, and representing problems. In J. E. Davidson & R. J. Sternberg (Eds.), *The psychology of problem solving* (Kindle Version). New York: Cambridge University Press.

Resnick, L. B. (1988). Treating mathematics as an ill-structured discipline. In R. I. Charles & E. A. Silver (Eds.), *The teaching and assessing of mathematical problem solving* (pp. 32–60). Hillsdale, NJ: Lawrence Erlbaum.

Richardson, W. (2010). *Blogs, wikis, podcasts, and other powerful web tools for classrooms* (3rd ed.). Thousand Oaks, CA: Corwin.

Sawada, T. (1997). Developing lesson plans. In J. P. Becker & S. Shimada (Eds.), *The open-ended approach: A new proposal for teaching mathematics* (pp. 23–35). Reston, VA: National Council of Teachers of Mathematics.]

Scheiber, R. (1987, March/April). The wit and wisdom of Grace Hopper. *The OCLC Newsletter*. Retrieved from http://www.cs.yale.edu/homes/tap/Files/hopper-wit.html.

Shimada, S. (1997). The significance of an open-ended approach. In J. P. Becker & S. Shimada (Eds.), *The open-ended approach: A new proposal for teaching mathematics*. Reston, VA: National Council of Teachers of Mathematics.

Sousa, D. (2008). *How the brain learns mathematics* (Kindle Version). Thousand Oaks, CA: Corwin.

Spinelli, J. (1990). *Maniac Magee*. New York: Little, Brown.

Stockman, A. (2015, January 10). How celebration is compromising young writers and one way you can change this. In *Brilliant or Insane*. Retrieved from http://www.brilliant-insane.com/2015/01/why-celebration-is-killing-young-writers-and-how-you-can-save-them.html

Stoner, D. A. (2004). *The effects of cooperative learning strategies on mathematics achievement among middle-grades students: A meta-analysis* (Doctoral dissertation, University of Georgia).

Sylwester, R. A. (1995). *Celebration of neurons: An educator's guide to the human brain.* Alexandria, VA: Association for Supervision and Curriculum Development.

Teachers Throwing Out Grades. [ca. 2015]. In *Facebook* [Group page]. Retrieved from https://www.facebook.com/groups/teachersthrowingoutgrades

United States Department of Agriculture [USDA]. (2013, August 20). *Pennsylvania mushroom production* [Press release]. Retrieved from http://www.nass.usda.gov/Statistics_by_State/Pennsylvania/Publications/Survey_Results/mushroom%20aug%202013.pdf

Vilson, J. (2012, October 15). *Engaging students in math* [Web log post]. Retrieved from http://www.edutopia.org/blog/engaging-students-in-math-jose-vilson

Wallis, W. A., & Roberts, H. V. (1962). *The nature of statistics.* New York: The Free Press.

Webb, N. (2005, July). *Web alignment tool (WAT) training manual.* Retrieved from http://wat.wceruw.org/Training%20Manual%202.1%20Draft%20091205.doc

Wong, K. Y. (2002, April). *Helping your students to become metacognitive in mathematics: A decade later.* Retrieved from http://math.nie.edu.sg/kywong/Metacognition%20Wong.pdf

Yatvin, J. (2012, August 17). *What schools need: Vigor instead of rigor* [Web log post]. Retrieved from http://www.washingtonpost.com/blogs/answer-sheet/post/what-schools-need-vigor-instead-of-rigor/2012/08/16/68be3d0c-e7fb-11e1–8487–64e4b2a79ba8_blog.html

Zaccaro, E., & Zaccaro, D. (2010). *Scammed by statistics.* Bellevue, IA: Hickory Grove Press.

Index

Figures are indicated by *f* following the page number.

A SAGE Company

Helping educators make the greatest impact

CORWIN HAS ONE MISSION: to enhance education through intentional professional learning.

We build long-term relationships with our authors, educators, clients, and associations who partner with us to develop and continuously improve the best evidence-based practices that establish and support lifelong learning.

Solutions you want. Experts you trust. Results you need.